The distribution of each species is given on the basis of the county system in use before 1974 with a few subdivisions as necessary.

Seaweeds of the British Isles

A collaborative project of the British Phycological Society
and the Natural History Museum with financial support from
the Joint Nature Conservation Committee

Volume 4

Tribophyceae (Xanthophyceae)

Tyge Christensen
University of Copenhagen

Natural History Museum, London

First published by the Natural History Museum,
Cromwell Road, London SW7 5BD

This edition printed and published by Pelagic Publishing, 2013,
in association with the Natural History Museum, London

ISBN 978-1-907807-73-2

This book is a reprint edition of 0-11-31004-3.

A catalogue record for this book is available from the British Library.

Seaweeds of the British Isles

Volume 1	Rhodophyta
Part 1	Introduction, Nemaliales, Gigartinales
Part 2A	Cryptonemiales *(sensu stricto)*, Palmariales, Rhodymeniales
Part 2B	Corallinales, Hildenbrandiales
Part 3A	Ceramiales
Part 3B*	Bangiophyceae
Volume 2	Chlorophyta
Volume 3	Fucophyceae (Phaeophyceae)
Part 1	Fucophyceae (Phaeophyceae)
Part 2*	Fucophyceae (Phaeophyceae)
Volume 4	Tribophyceae (Xanthophyceae)
Volume 5*	Cyanophyta
Volume 6*	Haptophyta
Volume 7*	Bacillariophyta

* To be published

Contents

Preface and acknowledgements

In writing this part of the seaweed flora I have largely followed the pattern established in the parts previously published. Some alterations in the format of the introductory text are due to the fact that this part deals solely with a single genus.

The material of this genus in public herbaria is rather scanty but comprises some interesting finds. I am indebted to the following institutions for permission to study either specimens from the British Isles or type material of species occurring in the British Isles:

Herbarium of the University of Michigan, Ann Arbor (USA) MICH
Botany School, University of Cambridge CGE
Botanical Museum, Copenhagen (Denmark) C
Royal Botanic Garden, Edinburgh E
Department of Botany, University of Glasgow GL
British Museum (Natural History), London BM
Botanical Museum, Lund (Sweden) LD
The New York Botanical Garden, New York (USA) NY
Department of Botany, University of Oxford OXF
Laboratoire de Cryptogamie, Muséum National d'Histoire Naturelle, Paris (France) PC

Very many colleagues have contributed by sending me living or preserved material, by giving valuable advice on suitable collecting sites, by helping me during my visits to their respective regions, or by offering important information. At the risk of omitting others who should have been mentioned, I wish to express my particular gratitude to Mr P. K. C. Austwick, Prof. J. L. Blum, Dr G. T. Boalch, Dr W. F. Farnham, Mr N. I. Hendey, Dr B. S. C. Leadbeater, Dr M. Parke, Mr H. T. Powell and Prof. A. Rieth.

My sincere thanks are due to Mrs Cynthia Gyldenholm for her careful revision of the English text.

Last, but not least, I thank the authors of the two preceding parts of the flora, Prof. P. S. Dixon and Mrs L. M. Irvine, for much help and advice in all matters.

TRIBOPHYCEAE Hibberd

TRIBOPHYCEAE Hibberd (1981), p. 95.

Xanthophyceae P. Allorge ex Fritsch (1935), p. 470,
non Xanthophyceen Correns (1893), p. 635.
Heterokontae A. Luther (1899), p. 17.
Numerous other synonyms, cf. Silva (1980).

Thalli of very diverse type. Chloroplasts green, containing chlorophyll a and variable amounts of chlorophyll c, but no chlorophyll b or fucoxanthin. Reserve substances fatty oil and sometimes chrysolaminarin, but never starch. Cell walls composed of mucilages and cellulose, the latter substance being masked by the mucilages so that there is no immediate colour reaction of the walls with zinc chloride and iodine. Monadoid cells usually (but not in the zoospores of *Vaucheria*) with one smooth and one hairy flagellum.

Most representatives of the Tribophyceae are normally found in fresh water or in soil, only exceptionally do they occur in slightly brackish water in salt marshes. Of the genera with regular marine or brackish water representatives, some are unicellular or colonial, and one is paucicellular and belongs in the plankton; only two are larger and bottom-living: *Tribonema,* which has multicellular filaments, and *Vaucheria,* which is siphonous. Each of the latter genera typify an order and a family name, the former also the name of the class.

While a species of *Tribonema* is known from the Mediterranean (Feldmann, 1941), this genus can no longer be regarded as a member of the British marine flora since the alga recorded under the name of *Tribonema endozooticum* (Wille) Magne has been shown to be a green alga (Christensen, 1985). Thus, the only tribophycean genus that falls within the framework of the present survey is *Vaucheria.*

VAUCHERIALES Bohlin

VAUCHERIALES Bohlin (1901), p. 14.

Heterosiphonales Pascher (1912), p. 21.

Thalli consisting of siphonous filaments with indefinite apical growth and lateral branching. Sexual reproduction oogamous. Gametes formed in parts separated from the vegetative thallus by double walls; several spermatozoids are formed in each antheridium and a single egg cell in each oogonium.

As conceived here, cf. Christensen (1980), the order only includes the family Vaucheriaceae.

VAUCHERIACEAE Dumortier

VAUCHERIACEAE Dumortier (1822), p. 71, 99.

Members of this family show considerable variation in the morphology of their sexual organs, but are rather uniform in the vegetative state, except for the existence of two types of chloroplasts. A separation of several genera, characterized by types of fruiting organs might be reasonable in principle but so far only a single step has been taken in such a direction, namely by the introduction of the generic name *Woroninia* by Solms-Laubach (1867) for the species otherwise called *Vaucheria dichotoma;* this has been rejected by later authors. One reason for not splitting the genus is the fact that some of the subdivisions currently used are rather badly founded; another is the convenience of having a single generic name for all sterile material of Vaucheriaceae met with in nature. Three other genera proposed within the family are probably based on misinterpretations: the separation of the genus *Vaucheriopsis* Heering (1921) was mainly based on what seems to be a wrong statement as to the storage product (see Whitford, 1943); the establishment of the genus *Pseudodichotomosiphon* Yamada (1934) was founded on vegetative features that may well have been the result of environmental factors (cf. Luther, 1953), and the type illustrations of the genus *Debsalga* Habeeb (1965) clearly show that its distinctive feature, the large supposed sporangium containing many spores, is a gall enclosing the eggs of the rotatorian *Proales wernecki,* as often found in various freshwater species of *Vaucheria. Ectosperma* Vaucher (1803) and *Vaucheriella* Gaillon (1833), the two additional synonyms of *Vaucheria* listed below, were only proposed as substitute names, not with the purpose of dividing the genus.

VAUCHERIA A. P. de Candolle

VAUCHERIA A. P. de Candolle (1801), p. 20.

Type species: *V. disperma* A. P. de Candolle (1801), p. 21 (= *V. canalicularis* (Linnaeus) Christensen).

Ectosperma Vaucher (1803), p. 25, nom. superfl.
Vaucheriella Gaillon (1833), p. 33, nom. superfl.
Woroninia Solms-Laubach (1867), p. 366.
Vaucheriopsis Heering (1921), p. 96.
Pseudodichotomosiphon Yamada (1934), p. 83.
Debsalga Habeeb (1965), p. 1, nom. illeg.

Vegetative structure

The width of the filaments usually lies between 12 and 200 μm according to the species. In most cases the width is approximately constant throughout a particular thallus. Some species, however, may produce downwardly growing rhizoidal filaments that are thinner; others form true rhizoids which are colourless and taper to a diameter that is a small fraction of the normal width. In species with terminal sporangia or gametangia the width often increases slightly towards the fertile tips.

The central part of the filament is occupied by a large vacuole. The parietal cytoplasm contains a layer of chloroplasts next to the wall. These are more or less elongate and are normally oriented parallel to the axis of the filament. In some species each chloroplast has a bulging pyrenoid. Such chloroplasts are approximately lanceolate and their ends may taper to a narrow point. Chloroplasts devoid of pyrenoids are generally elliptical or rounded-rhombic in outline. In the same layer there are usually lipid droplets. Most nuclei are located inside the chloroplast layer and are little revealed except in the occasional gaps between the chloroplasts. They are approximately spherical and generally smaller than the chloroplasts. The nuclear envelope persists throughout mitosis (Ott & Brown, 1972).

Elongation of the filament is exclusively by apical growth. Branching is always lateral (also in *Vaucheria dichotoma*). When submerged, the filaments usually grow towards the light and are relatively little branched. On moist ground, growth may be mainly decumbent and creeping with ample branching leading to the formation of felt-like coverings; however, where there is an unlimited supply of water and nutrients from below and adequate light such coverings gradually develop into velvet-like carpets or cushions of densely packed, erect filaments.

Reproduction

In addition to oospores, some species produce aplanospores; zoospores are only known in a few freshwater species. The formation of sexual organs often occurs when growth is slowed down by a diminution of water or nutrient supply. What induces the formation of aplanospores is not known.

Aplanosporangia and most antheridia develop from apical parts of filaments. The reproductive sector is separated from the vegetative thallus by two walls, one formed by the apical cytoplasm, the other by the cytoplasm below. In some species the two walls formed under a developing antheridium are rather far apart, enclosing a part of the filament which contains nothing but water, referred to as an empty space or, even more misleadingly, as an empty cell. Alternatively, the two walls may touch except for a narrow marginal part, giving the impression of a single transverse wall across the filament. During spermatozoid development, a pair of flagella is formed at each nucleus before cleavage of the cytoplasm and, at least in some species, even before separation of the fertile part from the vegetative thallus (Ott & Brown, 1978). The spermatozoids have no eyespots and no chloroplasts. They are heterokont, with a so-called proboscis supported by a broad microtubular root which originates near the base of the hairy flagellum (Moestrup, 1970). After their release a residue is seen in the antheridium containing, among other things, the relatively few chloroplasts of the fertile cytoplasm. The formation of aplanospores in the genus has not yet been studied in detail.

In some species antheridia and aplanosporangia develop from the tips of ordinary filaments, after which vegetative growth may continue sympodially with the formation of laterals below the fertile tips (Figs 1A, 1H, 4G, 4I, etc.). More usually, their formation is predominantly lateral, with additional terminal formation occurring now and then. In most species of the latter type the reproductive organs develop from the tips of special fruiting branches, which are short fertile shoots arising more or less perpendicular to the vegetative shoot. A smaller number of species produce lateral antheridia that are sessile on the vegetative filaments. In both cases there is often simultaneous formation of antheridia or sporangia along the parent filament, while the species with terminal antheridia and sympodial growth must necessarily form them in succession. Sympodial branching associated with the successive formation of antheridia is also common on fruiting branches and leads to the clustering of the sexual organs (Figs 3B, 4A); this phenomenon is referred to as proliferation. In the case of asexual sporangia, continued vegetative growth most usually takes place via the emptied sporangia (Fig. 2G). A complication of the process of antheridium formation is that of the separation of a part of the thallus which does not itself form spermatozoids but gives rise to a small number of antheridia, the parent, central part remaining sterile as a so-called androphore. Androphores are

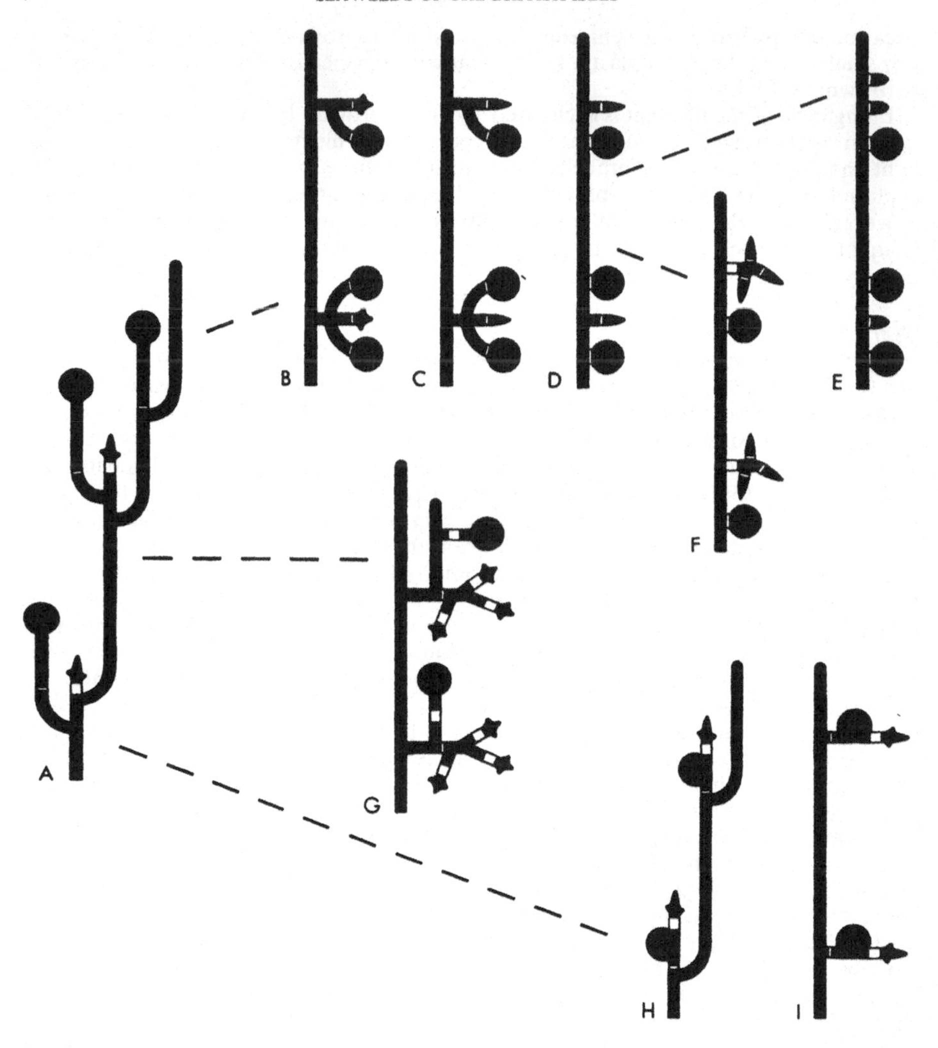

Fig. 1 Diagram illustrating the modes of branching at the sexual organs as found in British species of *Vaucheria*. The curving of many sexual organs is not shown. The species covered by the key on p. 8 fall within the various types as follows: A, *V. piloboloides;* B, *V. canalicularis* and *V. cruciata*; C, *V. erythrospora*; D, *V. arcassonensis*; E. *V. dichotoma, V. sescuplicaria* and *V. velutina*; F, *V. synandra*; G. *V. medusa*; H. *V. subsimplex*; I, *V. coronata, V. intermedia* and *V. minuta*. In two species the branching is simplified by dioecism, *V. compacta,* where the type is probably modified from I, and *V. litorea,* where it is probably modified from A or H. In *V. vipera* the branching type is doubtful. The lines indicate the possible derivation of one type from another but are not generally meant to suggest phylogenetic connections between the species mentioned.

formed in two British taxa; they are rather different in the two and are described in detail under the respective species.

The oogonia are separated from the vegetative thallus by a double wall like the antheridia, but only two British species show an empty, water-filled space between the two walls. Where the antheridium is terminal, either on the main filament or on a special fruiting branch, development of an oogonium usually starts with the formation of a bulge on the part of the filament just beneath the antheridium. In most cases a short branch is formed with the oogonium developing in a terminal position (Figs 1A–C, 6C), but in some species the oogonium develops directly from the bulging part of the filament so that its basal part becomes intercalated between the antheridium and the vegetative thallus (Figs 1H–I, 5F, 7C, etc.). Where the antheridium is sessile on the vegetative filament or terminates a very short fruiting branch, the oogonium starts as a bulge from the main filament and the mature oogonium becomes sessile, or almost so, on this filament (Figs 1D–E, 2A–C, 8A–B). The young oogonium contains numerous nuclei, but where the process of maturation has been followed, all but one of these disappear before fertilization. Two brackish-water species have been studied in this respect: in sections of *V. dichotoma* Heidinger (1908) found an accumulation of nuclei in the filament just below the oogonium, and Hansen (1972), using electron microscopy, found the same in the allied species, *V. sescuplicaria*. In living material of several freshwater species Heidinger and, earlier, Oltmanns (1895) observed a distinctive mass of protoplasm moving out of the oogonium just before its separation from the vegetative part of the thallus. In sections they found nuclei densely crowded along the same track. These observations lend support to the assumption that, in the species in question, the surplus nuclei are transferred back into the vegetative thallus before the separation of the oogonium. However, this may not apply to the genus as a whole: in *V. litorea* a considerable part of the protoplasm within the oogonium is separated from the apically located main portion and surrounded by a wall of its own (Fig. 4H); this part is often supposed to contain the surplus nuclei. In the allied species, *V. subsimplex*, a similar separation takes place within the oogonium, but the smaller basal part of the protoplasm usually dies without forming a wall. In other species related to the two an even smaller portion of protoplasm is left outside the egg, only occasionally with the formation of a wall around it (Fig. 6D), and in others nothing has been observed so far to suggest a separation of the surplus nuclei.

Sporangia are opened by a large apical pore. In certain cases the gametes are liberated by a gelatinization of the surrounding walls, but in the majority of species, including all those dealt with here, the sexual organs open by well-defined pores and fertilization takes place within the oogonium. In many species, pore formation is preceded by pronounced local swelling of the wall (Fig. 8B). The antheridium may empty through a more or less pointed apex, or through one or more lateral papillae, or both. The oogonium usually has a single pore, but in *V. coronata* several pores are formed (Fig. 3E).

The oospore is a resting organ, and is densely packed with oil droplets. Its chloroplasts may stay green, but more usually the green colour almost disappears during ripening, while reddish brown granules may form floccose accumulations at the periphery or a large mass in the centre. The wall of the oospore is usually thicker than that of the vegetative filament, often very considerably so. Thickened spore walls generally show obvious lamellation. In some cases three layers may be distinguished, and then the middle layer may be yellowish, brownish or reddish. Where no stratification is apparent the entire wall may take on a more or less marked yellowish or brownish hue.

Germination takes place by dehiscence of the outer wall layers at a preformed spot. At this place thicker walls often show a marked pit or groove in the outer layers (Fig. 5C), filled at maturity by a corresponding thickening of the innermost layer. This thickened part, which is not ruptured at germination, may be stained differentially with cotton blue (see Rieth 1980, p. 17).

It is generally assumed that meiosis takes place at germination so that the nuclei of the vegetative filaments are haploid, but the evidence for this is still weak.

Parasites and deformations

Fungal attacks are quite frequent in brackish water species of *Vaucheria. Phlyctochytrium vaucheriae* Rieth (1956a, p. 185), was found in *V. compacta, V. intermedia* and *V. velutina* by its author. In the British material it has been noted in the same species, as well as in *V. subsimplex*. Other parasitic phycomycetes are unidentified and perhaps undescribed.

On the inside of the filament wall there may be wart-like thickenings (Solms-Laubach, 1867) or finger-like ingrowths (P. A. Dangeard, 1925). The causes of these deformations are unknown.

Ecology

By its siphonous construction *Vaucheria* has effectively overcome the problem of water transport, which prevents most cellular algae from growing outside truly aquatic habitats. Also nutrients in the cell sap can simply diffuse from the parts of the thallus which have become buried in silt into the upper actively growing parts of the thallus, and protoplasmic elements from older parts may be shifted into those newly formed. Because of these properties species of *Vaucheria* compete very efficiently on mud in the intertidal zone in habitats where other upright filamentous algae would be impeded by desiccation at low tide and where creeping forms of algae suffer from the ever repeated deposition of silt. The free communication between emergent parts and parts covered by mud also enables some species to compete well on mud or sand that is constantly under water in sites where most other algae would suffer either from the effects of silting or from the lack of attachment.

Algae growing in the intertidal zone must necessarily tolerate very considerable changes in salinity at low tide; on the one hand heavy rain may dilute the salt water, on the other, strong insolation may result in concentrations above normal. One British species of *Vaucheria, V. piloboloides,* is marine and rather stenohaline, growing in the subtidal or just above low water mark. Two species, *V. canalicularis* and *V. cruciata,* are halophilous freshwater plants. When cultured in a stagnant medium they grow better in more or less brackish water than in fresh water, whereas, under natural conditions they are only occasionally found in brackish water but are common in running fresh water and still more so on the banks of streams and in other wet places. In the present work these halophilous freshwater species are not dealt with in detail although they are included in the key (p. 10). The remainder of the species dealt with tolerate 60‰ salinity as well as fresh water, at least for a period of time. Some, such as *V. dichotoma,* even exhibit some growth at both extremes while fruiting in a narrower range between; others die relatively quickly at one or both extremes; the majority show optimum growth somewhere between 10 and 25‰ salinity, and in all cases the range for good growth under laboratory conditions is comparatively wide (Christensen, 1987). In nature, where development also depends on competition, tolerance of zygotes to drought and other factors, many species grow over a narrower range of environmental conditions. Occurrence of the diverse species from the seaward edge to the upper levels of salt marshes and upstream in estuaries has been recorded in detail in the Netherlands by Simons (1975) and along the Dutch, German and Danish Wadden Sea by Polderman (1979, 1980, 1980a).

Most species show a more or less pronounced preference for certain seasons. The complex ecological background for this is poorly known, but the periodicity itself has been well studied in the Netherlands (Nienhuis & Simons, 1971; Simons, 1975a). The information given in the following account concerning times of fruiting is based on observations from the British collections; it is much more incomplete than the records of occurrence, because most samples have only been cursorily inspected at the time of collection, the species being subsequently determined in crude cultures.

Systematics and floristics

The Linnaean species of *Vaucheria,* all from fresh water except *V. dichotoma,* were originally only characterized by their vegetative features. Oogonia were first observed in a freshwater species by Müller (1779, cf. 1788) and then in *V. dichotoma* by Mertens, who left his material to Roth to describe. The latter author (1797) also saw antheridia but mistook these for younger stages of the oogonia or oospores, which he referred to as capsules. Blumenbach (1781) saw asexual reproduction, probably by aplanospores, Vaucher (1803) made a culture experiment with aplanospores and illustrated antheridia in several freshwater species, and Trentepohl (1807) reported swarming zoospores, but the understanding of these structures was poor. In following years, new species of *Vaucheria* were still being described without any record of antheridia, among these the brackish-water *V. litorea* C. Agardh (1823) and *V. velutina* C. Agardh (1824) and the marine *V. piloboloides* Thuret (1854). When Pringsheim (1855) recorded spermatozoids as well as describing the actual process of fertilization, attention was suddenly directed towards the male organs. Thuret (in Le Jolis 1863) made his description of *V. piloboloides* more complete by adding a text on the antheridium of this species, and in the monograph by Walz (1866) considerable taxonomic importance was attached to antheridial characters. Subsequently four additional species from brackish water were described: *V. subsimplex* by the Crouan brothers (1867), *V. synandra* by Woronin (1869), and *V. coronata* and *V. intermedia* by Nordstedt (1879).

After Götz (1897) had published a well-illustrated account of the *Vaucheria* species found in the region around Basel, central Europe, Heering (1907) provided a survey of all the species described until then. For more than thirty years this work was the standard reference work for botanists working on *Vaucheria.* It rendered good service as an unusually reliable guide to the pertinent literature but Heering knew the species mainly from the literature and herbarium specimens; he had collected little himself, and not at all from brackish water, as is stated in his monograph. Another drawback was that Heering did not recognize the principle of priority in nomenclature. Instead, he gave preference to names that were connected with good illustrations placing earlier names in synonymy, even where he had personally checked the type of an earlier name, or where the type had been studied by a contemporary specialist.

No further brackish water species were described from the North Atlantic area until Taylor (1937) established *V. compacta* referring to a manuscript left by Collins. Dangeard (1939) studied the *Vaucheria* flora of south-west France, giving original illustrations of all the species. Three of these were new to science and one of them, *V. arcassonensis,* was from brackish water. Regarding nomenclature, Dangeard followed the monograph by Heering. Works since 1950 by Christensen, Blum and Rieth have made several contributions to the *Vaucheria* flora along the North Atlantic, establishing the species *V. sescuplicaria, V. medusa, V. minuta, V. erythrospora* and *V. vipera.* In addition to their papers on more limited subjects, Blum (1972) has monographed the North American flora, and Rieth (1980), still adhering to the nomenclature of Heering, has surveyed all the freshwater species. A universal monograph by Venkataraman (1961), like its predecessor by Heering, is mainly based on literature but, unlike that classical work, is not very carefully done.

A few additional species of *Vaucheria* from brackish water have been described in papers dealing with the flora of geographical areas far from Britain. More pertinent to the context of the present work is the publication of further information on the species of the region and new illustrations of them. Espccially useful is the paper by Knutzen (1973) on the brackish-water species of southern Norway, Ott & Hommersand's (1974) account of the species of North Carolina, and Simons & Vroman's (1968) on finds from the Netherlands.

Papers dealing with brackish-water species of *Vaucheria* from the British Isles are relatively few and part of the early literature should be used with considerable caution, in particular with regard to the species that have not been very precisely defined in the past. Nordstedt's visit to Britain, after he had studied the genus in southern Sweden, led to the publication of two short papers, one by Holmes (1886), inspired by Nordstedt's observations, and another by Nordstedt (1886) with some corrections and supplementary information. After the turn of the century contributions were made by Cotton (1912), Carter (1933) and Chapman (1937), and in more recent years Christensen (1952, 1957), Cullinanc (1974), Polderman (1974, 1978) and Polderman & Polderman-Hall (1980) have recorded species new to the flora or added considerably to the number of known stations for more familiar species. Most of the information given in this flora, however, is based on previously unpublished finds, mainly collections made by the author, but also samples sent by other phycologists for identification and species loaned from public herbaria.

Remarks on the descriptions

For most species the average width of the filaments and the average dimensions of the oospores given are based on 100 measurements of material from the British Isles. In these cases the standard deviation is added. Although not statistically correct, the calculation of this value provides a tolerably good expression of the variation. The figures given for the thickness of the oospore wall are based on rather few measurements.

The illustrations are original, unless otherwise stated, and the original drawings were all made from material from the British Isles, cf. p. 35. They are reproduced at a magnification of 100 : 1, so that dimensions not given in the text can easily be measured by means of a millimetre rule.

Key to Species

1	Average vegetative filaments between 20 and 125 μm wide	2
	Filaments not falling within this range	26
2 (1)	With oogonia	3
	Without oogonia	20

3 (2) Oospore symmetrical about its longitudinal axis (spherical, ellipsoidal, obovoid or rounded lenticular), in most species not adhering to the wall of the oogonium.. 4

Oospore not symmetrical about an axis, adhering to the wall of the oogonium over most of its surface and usually flattened against its basal wall.. 16

4 (3) Oospore adhering to the entire wall of the oogonium or leaving only a narrow unfilled space at base and apex. Oogonium sessile on the vegetative filament and separated from it by a constriction usually less than half of the diameter of the filament .... *V. dichotoma* (p. 15)

Oospore not adhering to the wall of the oogonium except perhaps at the opening, in most species leaving wide unfilled space; if almost filling the oogonium then the latter separated from the vegetative filament by a constriction usually more than two-thirds of the diameter of the filament ... 5

5 (4) Oogonium terminal or forming a unilateral bulge below a terminal antheridium. A continuation of the vegetative filament may grow out laterally below the oogonium 6

Oogonium lateral to the main filament, either sessile or borne on a special fruiting branch or branch system that grows almost at a right angle from the main filament.. 8

6 (5) Plant dioecious. Oogonia and antheridia terminal. Female filament with oogonium hooked .... *V. litorea* (p. 19)

Plant monoecious.. 7

7 (6) Oogonium starting as a sack-like unilateral bulging of the filament immediately below a terminal antheridium, thus separating the latter from the vegetative filament. Oospore spherical....... *V. subsimplex* (p. 26)

Oogonium terminal on the main filament or on a branch that grows in the same direction, pear- to club-shaped, not separating the antheridium from the vegetative filament. Oospore rounded lenticular .... *V. piloboloides* (p. 23)

8 (5) Oogonium with several lateral openings from a broad apical papilla .... *V. coronata* (p. 14)

Oogonium with a single apical opening which may be difficult to observe in some species 9

9 (8) Oogonium sessile or borne on a short pedicel that does not bear antheridia 10

Oogonium borne together with one or more antheridia on a special fruiting branch system that grows from the main filament almost at a right angle.. 14

10 (9) Plant dioecious. Oogonium club-shaped. Antheridium spindle-shaped with cylindrical base .... *V. compacta* (p. 13)

Plant monoecious.. 11

11(10) Antheridium terminal on the main filament .... *V. vipera* (p. 30)

Antheridium lateral on the main filament or appearing as if in a lateral position on the oogonium.. 12

12(11) Antheridia sessile on the vegetative filament, usually forming a small group close to the oogonium. The latter pear-shaped, usually bent towards the antheridia.......................... *V. velutina* (p. 28)

Antheridia sessile or pedicellate, situated either right at the base of the oogonia or apparently on the latter. 13

13(12) Antheridia and oogonia shortly club- or spindle-shaped, sessile, separated from the filament by a narrow constriction, each with a single apical opening, one antheridium and one oogonium together on a slight elevation of the filament . . . . *V. sescuplicaria* (p. 25)

Antheridia tubular with two or more exit papillae, one of which is apical or nearly so, often apparently growing from the basal part of the almost spherical oogonium. . . . . *V. intermedia* (p. 19)

14 (9) Oogonium almost spherical. Antheridium usually on a short stalk . . . . *V. intermedia* (p. 19)

Oogonium pear-shaped or shortly club-shaped. Antheridium usually on a long stalk 15

15(14) Antheridia usually 2–4 together, borne on a bottle-shaped structure that is terminal on a fruiting branch . . . . *V. medusa* (p. 20)

Each antheridium usually terminating a fruiting branch. . . . . . . *V. vipera* (p. 30)

16 (3) Oogonia sessile on the vegetative filaments or single on short pedicels 17

One oogonium or more borne together with one antheridium or more on a special fruiting branch that grows from the main filament almost at a right angle. In addition there may sometimes be a terminal antheridium with a lateral oogonium or more just below. 18

17(16) Oogonium only moderately asymmetrical, its largest diameter less than twice the diameter at the base; antheridium not borne on a special carrying structure. . . . . *V. arcassonensis* (p. 11)

Oogonium strongly asymmetrical with a hooked beak turned towards the antheridia, its largest diameter more than twice the diameter at the basal constriction; several antheridia borne together on an ellipsoidal structure, which is separated from the filament by two walls with an empty space between them . . . . *V. synandra* (p. 26)

18(16) Oogonium strongly asymmetrical with a hooked beak not filled by the oospore . . . . *V. erythrospora* (p. 17)

Oogonium only slightly asymmetrical, rounded, entirely filled by the oospore 19

19(18) Fruiting branch hooked under the antheridium. Vegetative filaments more than 40 μm wide. Halophilous freshwater species (not further dealt with here). . . . . *V. canalicularis*

Fruiting branch straight or only slightly bent under the antheridium. Vegetative filaments less than 35 μm wide. Halophilous freshwater species (not further dealt with here) . . . . *V. cruciata*

20 (2) With antheridia 21

Without antheridia 24

21(20) All antheridia young, none empty . . . Possibly monoecious plants that have not yet formed oogonia – see illustrations of both monoecious and dioecious species for determination

Some antheridia emptied 22

22(21) Antheridium terminal on the main filament. A continuation of the filament may grow out laterally below the antheridium . . . . *V. litorea* (p. 19)

Antheridia sessile or pedicellate in a lateral position on the main filament. . . .23

23(22) Antheridia spindle-shaped with cylindrical bases, usually each with several openings, single or with new pedicels growing laterally from older pedicels .... *V. compacta* (p. 13)
Antheridia clavate or shortly spindle-shaped with only a terminal opening, single or in clusters, each antheridium growing directly from the vegetative filament................................ *V. dichotoma* (p. 15)
24(20) Without aplanosporangia ... not determinable by direct microscopic observations
With aplanosporangia.. 25
25(24) Without pyrenoids. Filaments 21–29 μm wide .... *V. intermedia* (p. 19)
With pyrenoids – see *V. compacta, V. piloboloides, V. subsimplex* and *V. velutina*
26(1) Filaments less than 20 μm wide .... *V. minuta* (p. 22)
Filaments more than 125 μm wide...................... *V. dichotoma* (p. 15)

The Species

Vaucheria arcassonensis P. J. L. Dangeard (1939), p. 220.

Lectotype: original description, in absence of material. France (Arès).

Plants usually mixed with other species, forming mats or carpets on soil. Filaments 48 ± 6 μm wide, sometimes with long, branched rhizoids descending vertically from creeping shoots. Chloroplasts elongate, rounded or pointed at their ends, without pyrenoids.

Antheridium borne on a relatively short pedicel, tubular, a little narrower than the vegetative filament, irregularly bent or twisted, opening through an apical pore. Oogonium either borne on a short pedicel or sessile on the filament, shortly tubular, a little wider than the vegetative filament with its maximum girth above the middle, more or less bent with an inconspicuous apical beak pointing in the direction of the curvature, opening by a pore at the tip of this beak. Oogonia starting into growth soon after the antheridia and developing almost at the same time, usually one or two oogonia next to a single antheridium, often remarkably close to the tip of the filament. Oospore 115 ± 16 μm long, 80 ± 9 μm across, with a colourless wall about 5 μm thick; contents at maturity greyish with a reddish brown spot at the centre.

Asexual spores unknown.

Capable of growing between 2·5 and 50‰ salinity. Fruiting between 10 and 40‰, most abundantly between 15 and 30‰. In salt marshes, most usually under *Halimione,* but also often under *Spartina* and sometimes under other phanerogams. Generally mixed with other *Vaucheria* species, commonly with *V. intermedia* and quite often with *V. velutina* and *V. coronata*.

Recorded as scattered finds from all areas of the British Isles and probably more frequent than indicated by these finds, particularly as it seems to be most abundant in the colder months during which there has been relatively little collecting.

Denmark to south-west France, Maine to North Carolina.

British fruiting material has only been collected in May; sterile samples are recorded for most of the year but are infrequent during summer months.

Fig. 2 A–C. *Vaucheria arcassonensis*; D–G. *Vaucheria compacta*; D. Germinating aplanospore; E. Filament with antheridia; F. Filament with oogonium; G. Filament with one emptied and one newly formed aplanosporangium. All × 100.

Vaucheria compacta (Collins) Collins ex Taylor (1937), p. 226.

Lectotype: C. USA (Malden, Mass.). The type material is the specimen of Phycotheca Boreali-Americana no. 477 used for the published illustrations (see Christensen, 1952); all other specimens with this number are isotypes.

Vaucheria piloboloides Thuret var. *compacta* Collins (1900), p. 13.
Vaucheria sphaerospora Nordstedt var. *dioica* Rosenvinge (1879), p. 190.

Plants usually forming compact carpets on land, or looser masses of sparingly branched filaments growing vertically from the bottom of temporary, shallow water, often individually, but also quite frequently growing with other species. Filaments 38 ± 7 µm wide, on average a little wider in plants with asexual spores than in those with sexual organs, in the former gradually increasing towards the sporangia. Chloroplasts lanceolate with pyrenoids.

Sexual organs terminal on short fruiting branches that grow at right angles from the vegetative filaments, sometimes almost sessile, often several at the same stage of development on one filament. Male and and female organs normally on separate plants (exceptional finds of monoecious plants are reported from the Netherlands by Simons, 1974). Antheridium separated from the vegetative part of the fruiting branch by an empty space, elongate, straight, increasing slightly in width from its base towards the middle part, apically tapering into a point, where an exit pore is always formed. Lateral exit pores variable in number and shape, most usually two pointing in opposite directions, one a little closer to the apex than the other. In salt and brackish water the exit pores are normally situated on rather large conical projections, in fresh water they are often less protruding. Proliferation from the antheridium stalk frequent, rather similar to that seen in *V. coronata* (Fig. 3B). Oogonium club-shaped, opening at the apex without the preceding formation of a papilla, so that the opening is barely obvious in lateral view. Contents of the oogonium contracting into an almost spherical egg with no, or very little, protoplasm abandoned outside it. Oospore spherical or slightly greater in length than breadth, length 125 ± 13 µm, breadth 123 ± 12 µm; wall about 3 µm thick, colourless or pale yellow; contents colourless at maturity.

Asexual reproduction by thin-walled, obovoid aplanospores formed terminally on vegetative filaments or short fruiting branches. After shedding, the parent filament often continues growth through the base of the emptied sporangium.

The species shows some morphological variation: Blum and Wilce (1958, 1958a) describe a variety from Canada with unusually long oogonia; from the Netherlands Simons (1974) describes material with a lower than average number of exit pores on the antheridia, typically developed in fresh water, and Dangeard (1939) draws attention to plants with various types of aplanosporangia occurring along the major rivers of France.

Grows tolerably well up to a salinity of 50‰ and is also capable of growing in fresh water. In fresh water it only occurs by rivers. Under brackish water conditions it also grows where there is no water movement, as in salt pans, but it is especially common, and often entirely dominant, in the intertidal zone in the parts of estuaries where brackish and fresh water alternate, and on beaches where fresh water oozes out from the ground associated with regular flooding with brackish water at high tide, conditions which most other species do not tolerate so well. In such sites *V. compacta* often forms extensive carpets of densely compacted, upright filaments, a growth habit alluded to by the specific name but not restricted to this species. In addition, the species is commonly found in salt marshes under various phanerogams and on the exposed sides of creeks, and is then often mixed with other species, such as *V. intermedia, V. velutina* and *V. subsimplex*. Simons (1974) gives a very detailed account of the ecology of *V. compacta* in the Netherlands.

Many finds from England, fewer from Scotland and Ireland.

Southern Norway to south-west France, Labrador to the Gulf of Mexico.

Fruits from late spring to autumn, mainly in July and August. Aplanospores often formed under cultural conditions but rarely found together with sexual organs in nature; this combination was observed in August.

Vaucheria coronata Nordstedt (1879), p. 177.

Lectotype: LD. Sweden (Landskrona). The type material is a mixture of *V. coronata* and *V. intermedia* Nordstedt. It is no. 344 of Wittrock & Nordstedt: Algæ aquæ dulcis exsiccatæ, all other specimens with this number being isotypes.

Plants most usually forming a relatively thin covering on soil, generally rather loosely tangled, sometimes more densely matted when growing with other species. Filaments 39 ± 4 µm wide. Chloroplasts elongate, usually rounded-hexagonal, without pyrenoids.

Male and female organs generally borne together on fruiting branches, more rarely they develop from part of the main filament. In principle, branching at the sexual organs is of the type shown in Fig. 1I, but in most cases proliferation takes place from below the first antheridium before the oogonium is initiated (Fig. 3B) and the oogonium then usually develops from below the empty space first formed. Not uncommonly, a third antheridium is formed in the primary or the secondary fruiting branch from the part that becomes the physiological apex after separation of the first formed antheridium (Fig. 3D). The tip of a vegetative filament may develop into an antheridium like the tip of a fruiting branch and a seemingly intercalary antheridium may arise in the same way behind an inner, apparently transverse wall formed towards a moribund part of the filament. Such seemingly intercalary antheridia in fruiting branches or main filaments are virtually unknown in all other British species. No matter where formed, the antheridium is tubular, as wide as the vegetative filament or nearly so, and separated from the vegetative part by an empty space. Opening usually takes place through a single short and broad exit tube which may be situated at the middle of the antheridium, perpendicular to its longitudinal axis, or close to the distal end and then often directed obliquely upwards. The oogonium usually develops from a lateral bulge of the fruiting branch close to its empty part, more rarely from a similar bulge of a main filament below an antheridium or a seemingly transverse wall. It is shortly ovoidal, laterally attached and crowned by a circle of short excrescences, the tips of which eventually dissolve, providing a passage for the sperm cells. Oospore approximately spherical, 111 ± 13 µm long, 107 ± 10 µm across; wall at maturity about 5 µm thick, yellowish brown, often with a delicate marking; contents colourless at maturity or with a rusty central spot.

Asexual spores unknown.

Growing tolerably well up to a salinity of 45‰ and also capable of growing in fresh water in the laboratory, although never occurring in fresh water in nature. Fruiting at salinities between 2·5 and 30‰. Found in salt marshes, most usually under *Halimione*, but also often in other vegetation types, very often mixed with *V. intermedia* and not infrequently with *V. arcassonensis, V. compacta, V. synandra,* and *V. velutina.*

Generally distributed throughout the British Isles.

Greenland to south-west France and North Carolina.

Sterile material found throughout the year. Most fruiting samples are from the spring.

Vaucheria dichotoma (Linnaeus) Martius (1817), p. 304.

Type: the description by Linnaeus, based on that by Dillenius (1742), which in turn is based on material kept in the *Historia muscorum* herbarium at OXF (cf. Christensen, 1968).

Conferva dichotoma Linnaeus (1753), p. 1165.
Vaucheria pyrifera Kützing (1843), p. 305.
Vaucheria starmachii Kadłubowska in Starmach (1972), p. 587, nom. nud.

Plants usually forming bearskin-like masses in calm water. Dimensions extremely variable (cf. Rieth, 1978), frequently surpassing those of all other species. Chloroplasts without pyrenoids, very variable in shape and size, usually elliptical or almost circular in outline.

Antheridium ellipsoidal to rounded-fusiform with a small apical papilla, sessile on the main filament; several organs of different age often clustered together. Oogonium usually a little more than twice as long as the antheridium, ellipsoidal, shortly obovoid or almost spherical, likewise with an apical papilla; after it has reached full size but before it is separated from the filament most of its cytoplasm and almost all its chloroplasts collect in the apical half, forming a very sharp boundary towards the nearly colourless basal half. Both organs open by dissolution of the apical papilla. The oospore usually entirely fills the oogonium, but may occasionally leave a narrow open gap at its base and apex. Its wall is about 5 µm thick at maturity, smooth, pale brown to light golden brown. The contents are almost colourless, with a few green chloroplasts and a small proportion of golden-brown granules between masses of oil droplets. Antheridia and oogonia may be formed on the same filament and are then scattered with no mutual relationship as to position; they may also be separate so that some parts carry only antheridia while other parts of the same thallus bear only oogonia; one clone from continental Europe was found to form both kinds of sexual organs in one culture vessel and only one in another under seemingly identical conditions, and other clones were strictly unisexual in culture.

Asexual spores unknown.

Grows reasonably well in fresh water and tolerably at a salinity of 40‰ but most clones fruit little or not at all outside salinities between 5 and 15‰. The type locality in Kent, now considerably disturbed, agreed well with the majority of stations for this species known from other parts of the world – freshwater localities which receive some salt or brackish water now and then. In fresh water the species often shows luxuriant growth but rarely fruits. It may cause drainage trouble in lodes and ditches (see Dowidar & Robson, 1972), and may also cover the muddy bottom of lakes to a considerable depth.

What entities are covered by the names *Vaucheria dichotoma* var. *submarina* Lyngbye (1819), p. 76 and *Vaucheria submarina* (Lyngbye) Berkeley (1833), p. 24, both typified by Lyngbye's description and illustrations, remains doubtful. The plant from Weymouth

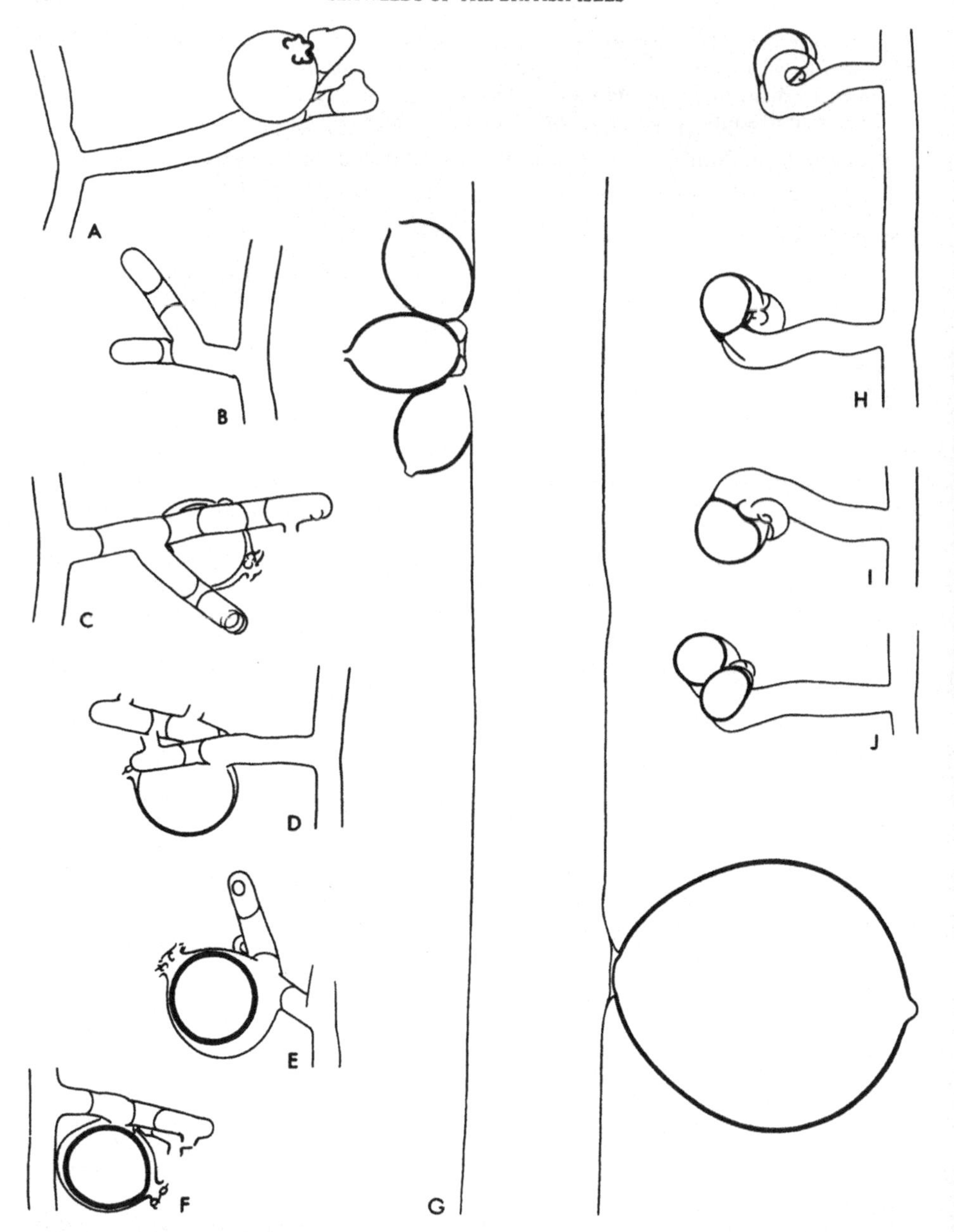

Fig. 3 A–F. *Vaucheria coronata*; G. *Vaucheria dichotoma*; H–J. *Vaucheria erythrospora*. All ×100.

referred to this species or variety by Berkeley (1833) and Holmes (1886) and that from Norfolk referred to the variety by Chapman (1937) are hardly identical. That from Weymouth may be *V. velutina* var. *separata* (see p. 30).

Few brackish water finds are recorded, principally from south-east England and the west coast of Ireland.

Norway to Algeria, Canada, Bermuda, Antilles.

Fruiting material known from May, June, September and October.

This is the only species which has been grown in axenic culture. It was isolated by Åberg & Fries, who in 1976 described its growth at different temperatures and in different media; subsequently Åberg (1978) tested the effect of light and wounding on branch formation.

Vaucheria erythrospora Christensen (1956), p. 275.

Lectotype: original description by Rieth (see below), in the absence of material. Germany (Artern).

Vaucheria hamata [sensu Götz] f. *salina* Rieth (1956), p. 139.

Plants most usually forming a relatively thin cover of loosely tangled filaments on soil. Filaments 38 ± 4 µm wide, sometimes forming rhizoids. Chloroplasts ellipitical to almost circular plates without pyrenoids.

Male and female organs borne together on a system of fruiting branches. Antheridium tubular, terminal on the fruiting branch of the first order, coiled together with the distal part of the branch, opening at the apex. Oogonium rounded at the base and much broader than the supporting filament, pointed at the apex, terminal on a shorter fruiting branch of the second order, one or two such branches arising from the branch of the first order approximately where its coiling begins; both the oogonium and the branch that carries it curved so that the beak of the oogonium comes close to the opening of the antheridium. Opening of the oogonium takes place at the tip. The zygote rounds off at the apex so that the beak of the oogonium is left empty, while the rest of the oospore wall adheres to the wall of the oogonium. The oospore is 85 ± 7 µm long and 67 ± 5 µm wide in lateral view. At maturity the wall is about 3 µm thick, made up of two colourless layers and a median, thinner, red layer. In the centre of the spore there is a reddish brown spot; the rest of the contents are colourless.

Asexual spores unknown.

Grows tolerably well up to a salinity of 45‰ and is capable of growth in fresh water. Fruiting occurs over the range from 0 to 35‰. In nature never occurring in fresh water and rarely at higher salinities. Normally found in the upper parts of salt marshes, where the salinity is usually low and the surface soil often rather dry. Sometimes found unmixed, sometimes mixed with other species, in particular, *V. sescuplicaria*.

British finds only from south-east England.

Denmark to the Netherlands. Scattered finds from North Africa, Atlantic coast of North America, Pacific coast of Asia and south coast of Australia.

V. erythrospora seems to pass the summer as zygotes, which may be partly the reason why the species is unrecorded for most of the area. Fruiting material has been collected at all other seasons in neighbouring parts of Europe, but so far has not been found in Britain.

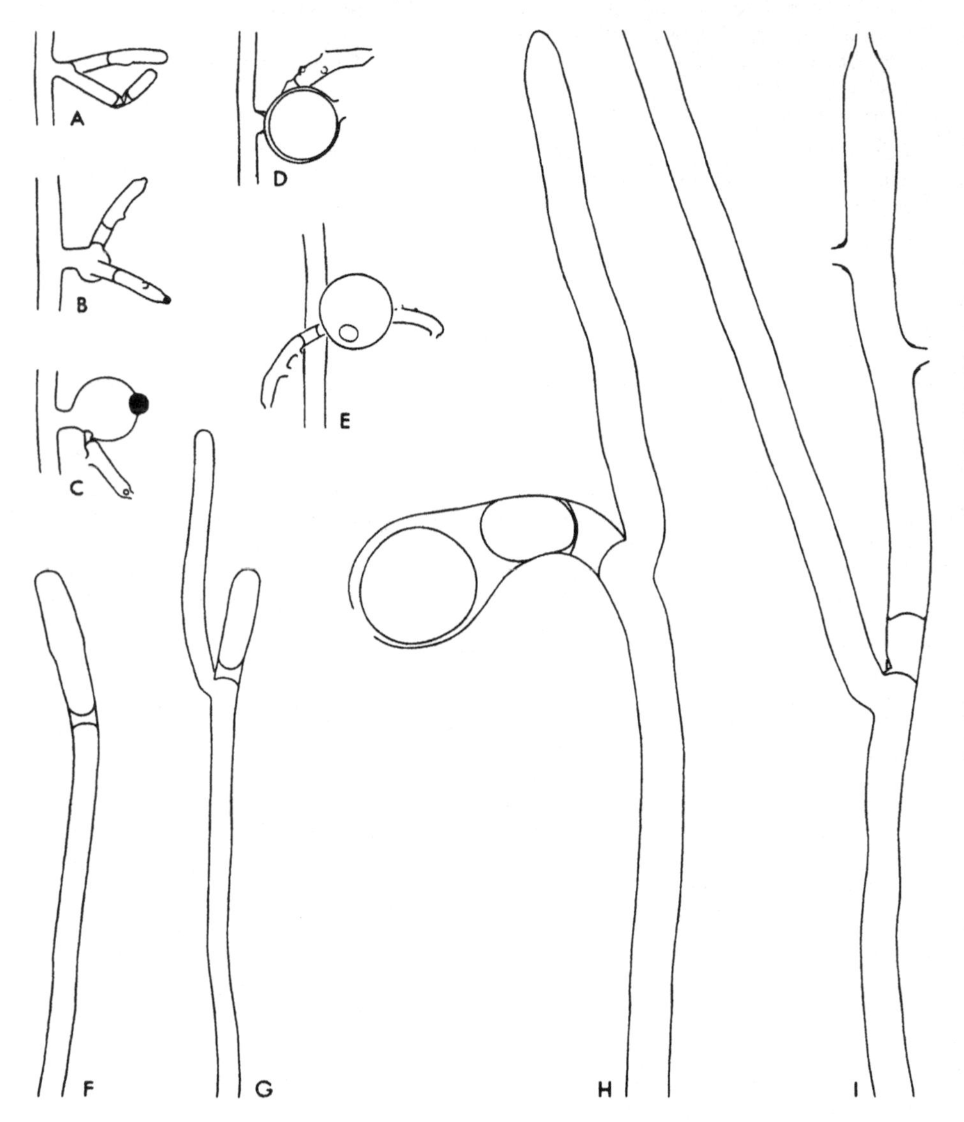

Fig. 4 A–G. *Vaucheria intermedia*; A–E. Sexual organs; F–G. Aplanosporangia; H–I. *Vaucheria litorea*; H. Oogonium; I. Antheridium. All × 100.

Vaucheria intermedia Nordstedt (1879), p. 179.

Lectotype: LD. Sweden (Landskrona). The type material is a mixture of *V. intermedia* and *V. coronata* Nordstedt (see p. 14).

Plants most usually forming rather dense carpets on soil. Filaments 25 ± 4 µm thick. Chloroplasts elongate–hexagonal with more or less rounded sides, without pyrenoids.

Male and female organs borne together on fruiting branches, which may be so short that eventually the oogonia appear to be sessile on the main filaments. Branching is of the type shown in Fig. 1I. Very often, however, proliferation takes place from the antheridium stalk before an oogonium is formed (Fig. 4A). The antheridium is tubular, a little narrower than the vegetative filaments, usually opening through a small, more or less excentric exit tube at the tip and a varying number of similar lateral tubes, most frequently one or two. The oogonium is almost spherical. It develops from the antheridium stalk immediately below the empty space, and by this process the antheridium is pushed aside so that it takes up a seemingly lateral position on the basal part of the oogonium. At the apex of the oogonium a rather large exit papilla is formed which dissolves at maturity leaving a well-marked prominent rim (Fig. 5D). Oospore spherical, 91 ± 12 µm in diameter, with a colourless wall, about 3 µm thick; contents dark green, even at maturity.

Asexual reproduction by almost cylindrical aplanospores, which are formed at the tips of ordinary filaments, usually not filling the whole length of the sporangia.

Capable of growing in fresh water and up to a salinity of about 45‰. Fruiting at salinities between 2·5 and 30‰. In nature, so far never found in fresh water but occurring at very diverse salinities in salt marshes. Common under *Halimione, Puccinellia* and *Aster* and on the steep sides of creeks, frequently mixed with other species, such as *V. compacta, V. velutina, V. arcassonensis, V. coronata* and *V. subsimplex,* but also often in pure stands.

Generally distributed throughout the British Isles.

Greenland to south-west France and Virginia, Washington State.

Occurs almost throughout the year. Fruiting material only collected in summer and autumn in Britain, but in the Netherlands also found in the winter. Asexual spores not found in nature in Britain but have been obtained in a crude culture.

Vaucheria litorea C. Agardh (1823), p. 463.

Holotype: C. Denmark (Hofmansgave, Funen). The species has traditionally been attributed to 'Hofman ex C. Agardh', but Hofman suggested the epithet for a different species (Christensen, 1986a).

Vaucheria clavata sensu Lyngbye (1819), p. 78, non *Vaucheria clavata* (Vaucher) A. P. de Candolle in Lamarck & De Candolle (1805), p. 63.

Plants sometimes growing submerged as masses of relatively long, sparsely branched, erect filaments, otherwise most usually forming dense carpets or irregular tufted stands, also upright growing. There is considerable variation in size (cf. Christensen, 1986a): The samples used for Fig. 4H and I have relatively small dimensions, measurement of 50 filaments from the two showing a width of 56 ± 4 µm. These samples were taken relatively high up in salt marshes, one in Glamorgan, the other in Carmarthen, and brought into fruit in crude cultures. The opposite extreme to the two samples just mentioned is represented by a plant that fruited submerged or recently detached in a Cornish boating lake connected to an estuary from about mid-tide, the same number of filaments from this material showing a width of 89 ± 8 µm. The filaments are often bluntly pointed at the tips, while all the other species dealt with here are hemispherically rounded. Chloroplasts highly variable in size and shape, some of them tapering to a narrow point, all with pyrenoids.

Sexual organs terminal on the main filaments, male and female organs normally on separate plants. Vegetative growth continuing sympodially from beneath the sexual organs as in Fig. 1H. Antheridium almost tubular, a little wider than the vegetative filament, separated from the latter by an empty space. Opening takes place through an apical exit tube and a varying number of similar lateral tubes, most usually two or three, all on small conical projections. Oogonium club-shaped, likewise separated from the vegetative filament by an empty space, usually curving together with the uppermost part of the filament so as to point obliquely downwards. Contents of the oogonium separating into a spherical egg, which fills the distal part, and a smaller sterile portion, which fills the proximal part and surrounds itself with a wall. The sterile portion, which probably contains all the nuclei except the single gamete nucleus, may occasionally grow out into a vegetative filament but usually dies after some time. Opening of the oogonium takes place by dissolution of the apical part of the wall without preceding formation of a papilla. Oospore approximately spherical, varying in dimensions like the vegetative filaments. 25 spores from the two salt marsh samples from Wales mentioned above measured 168 ± 16 µm in length and 165 ± 16 µm in width, while the corresponding figures for the Cornish boating lake sample were 243 ± 12 and 232 ± 12 µm. Spore wall about 7 µm thick, pale golden brown; contents almost colourless.

Asexual spores unknown.

Capable of growing in fresh water and up to a salinity of about 50‰. Fruiting at salinities between 2·5 and 40‰. In nature, never found where the water is constantly fresh but not unusually within reach of fresh water, preferably in very sheltered places, between reeds, in ditches and creeks, etc., often mixed with *V. compacta* and sometimes also *V. intermedia*. One sample was taken in a fully marine environment, covering the sand between rocks with fucoids at low-water mark.

Scattered finds from all parts of the British Isles.

Iceland to south-west France, Massachusetts to Louisiana, Washington State, Australia.

British finds from spring and summer, fruiting material only from April and May; in Denmark fruiting throughout the summer and until October.

Vaucheria medusa Christensen (1952), p. 179.

Holotype: C. Sweden (Trolle-Ljungby, Skåne)

Plants often creeping on soil but also sometimes occurring as dense carpets or cushions or buried in silt with only the filament tips protruding, more rarely as short, upright filaments on the floor of shallow water. Filaments 36 ± 6 µm wide. Chloroplasts lanceolate, with pyrenoids.

Male and female organs usually borne together on special fruiting branches. The primary fruiting branch has the same width as the main filament and may sometimes be

Fig. 5 A–C. *Vaucheria medusa*; D–G. *Vaucheria minuta*. All × 100.

relatively long. Its apical part is cut off by a double wall to become a so-called androphore. This special structure gives off a number of irregularly winding, thinner filaments, partly from its distal and partly from its proximal end; each of these thinner branches then forms a terminal antheridium separated from the vegetative part by an empty space. The antheridium is more or less spindle-shaped with a terminal and, usually, 1–3 lateral exit tubes, each on a conical projection. The oogonium is also separated from the vegetative thallus by an empty space; it is club-shaped and recurved, terminal on a straight or curved branch. The branch may be a lateral of the second order growing from the primary fruiting branch just below the double wall (Fig. 1G below, Fig. 5C), or it may grow from such a lateral as a lateral of the third order (Fig. 1G above, Fig. 5A); more rarely it is a lateral of the first order grown from the main filament (Fig. 5B). The contents of the oogonium divide into a subspherical egg that fills the distal part and a relatively small sterile mass of protoplasm that is located in the proximal part; the sterile part occasionally surrounds itself with a wall but more often dies without forming one. Opening of the oogonium takes place by dissolution of the distal part of the wall without prior formation of a papilla. Oospore approximately spherical, 114 ± 7 µm long and 124 ± 9 µm wide, thus the breadth is often a little greater than the length; wall about 4 µm thick, smooth and colourless; contents almost colourless, with a few scattered chloroplasts and some reddish brown granules between the oil droplets.

Asexual spores unknown.

Capable of growing in fresh water and up to 20‰ salinity. Fruiting in culture between 5 and 15‰. In nature hardly to be found in absolutely fresh water but preferably on river banks in areas only reached by brackish water at the highest tides, often under freshwater phanerogams and generally associated with freshwater species of *Vaucheria* as well as brackish-water species, mainly *V. synandra, compacta, bursata, cruciata, canalicularis* and *frigida*. More rarely found in salt marshes behind embankments, etc.

Scattered finds from all over the British Isles.

Southern Norway to the Netherlands.

British finds from early summer to late autumn. Fruiting material from June.

Vaucheria minuta Blum & Conover (1953), p. 399.

Holotype: MICH. USA (Falmouth, Mass.), cf. Christensen (1986a).

Plants usually mixed with other species, forming carpets on soil. Filaments 17 ± 4 µm wide. Chloroplasts elongate, often bluntly pointed at the ends, without pyrenoids.

Male and female organs borne together on fruiting branches in the manner shown in Fig. 1I. Fruiting branches relatively long (unlike those in *V. intermedia,* which has the same mode of branching) and almost as wide as the main filament. Antheridium tubular, terminal on the fruiting branch, separated from it by an empty space, rounded at the apex, opening through a lateral, broad and flat exit papilla, the rim of which only slightly projects after dissolution of the main part. Normally without proliferation. Oogonium originating from a swelling of the fruiting branch just below the empty space, elongate with its longitudinal axis nearly continuing the direction of the fruiting branch, thus displacing the antheridium and the empty part of the tube to one side. The oogonium may have its basal wall at the beginning of the swelling or lower down, so that an unswollen tubular part of the fruiting branch is included. Apex conical with a large terminal pore.

Oospore ellipsoidal, 55±7 µm long, 44±5 µm in diameter, with a smooth, colourless wall usually 1–3 µm thick; contents greyish except for an orange-brown spot in the centre, more rarely two or three such spots are present.

The species shows some variation. The dimensions given above are based on an English and an Irish sample, 50 measurements taken from each. Both samples have smaller spores than the type material and Dutch material studied by Simons & Vroman (1968), while the Irish material has wider filaments than all the other collections. In material from North Carolina, studied by Ott & Hommersand (1974), ripe oospores were found to be dark green.

Asexual spores unknown.

Capable of growing between 2·5 and 50‰ salinity and fruiting between 10 and 20‰. Found in open patches between salt-marsh phanerogams, *Halimione, Puccinellia, Armeria,* etc. associated with other *Vaucheria* species, such as *V. coronata, V. arcassonensis* and *V. intermedia.*

Only known from Lincolnshire, Argyll, Wigtown, Hampshire, Dorset and Dublin. Denmark to the Netherlands, Maine to North Carolina.

Only collected in winter and spring; British fruiting material only recorded for April.

Vaucheria piloboloides Thuret (1854), p. 389.

Holotype: PC. France (Saint-Vaast-la-Hougue, Normandy).

Vaucheria fuscescens Kützing (1856), p. 20.

Plants usually growing submerged as relatively long, sparsely branched, upright filaments about 56 µm wide. Chloroplasts more or less lanceolate, each with a pyrenoid.

Fruiting organs terminal on vegetative filaments, but often more or less clustered as a result of the formation of short branches below the sexual organ first formed (Fig. 1A). Initial fruiting organ usually an antheridium but also occasionally an oogonium; sex of subsequent organs varying without any obvious regular sequence. Antheridium separated from the filament by an empty space, straight, tubular, over the greater part about as wide as the vegetative filament, tapering conically towards the tip, where there is always an exit pore. Lateral pores varying in number and shape, most commonly 1–3 situated at the tips of often rather long, tubular or conical projections. Oogonium club-shaped with a remarkably sudden transition from the tubular proximal part, which is often relatively long, to the inflated distal part, opening by the dissolving of part of the apical wall without prior formation of a papilla. Contents of the oogonium separating into an egg situated against the apical wall, and a smaller proximal part that may surround itself with a wall but usually dies without having formed one. Oospore, like the egg, more or less lenticular, about 135 µm long and 171 µm in diameter, adhering by its distal face to the inner layer of the oogonium wall; this layer often separating from the outer layer, which swells considerably with age. Spore wall about 6 µm thick, almost colourless; contents greyish.

Asexual reproduction by aplanospores formed in terminal sporangia which are more or less club-shaped but far less inflated than the oogonia. The spores may be unwalled at the time of liberation.

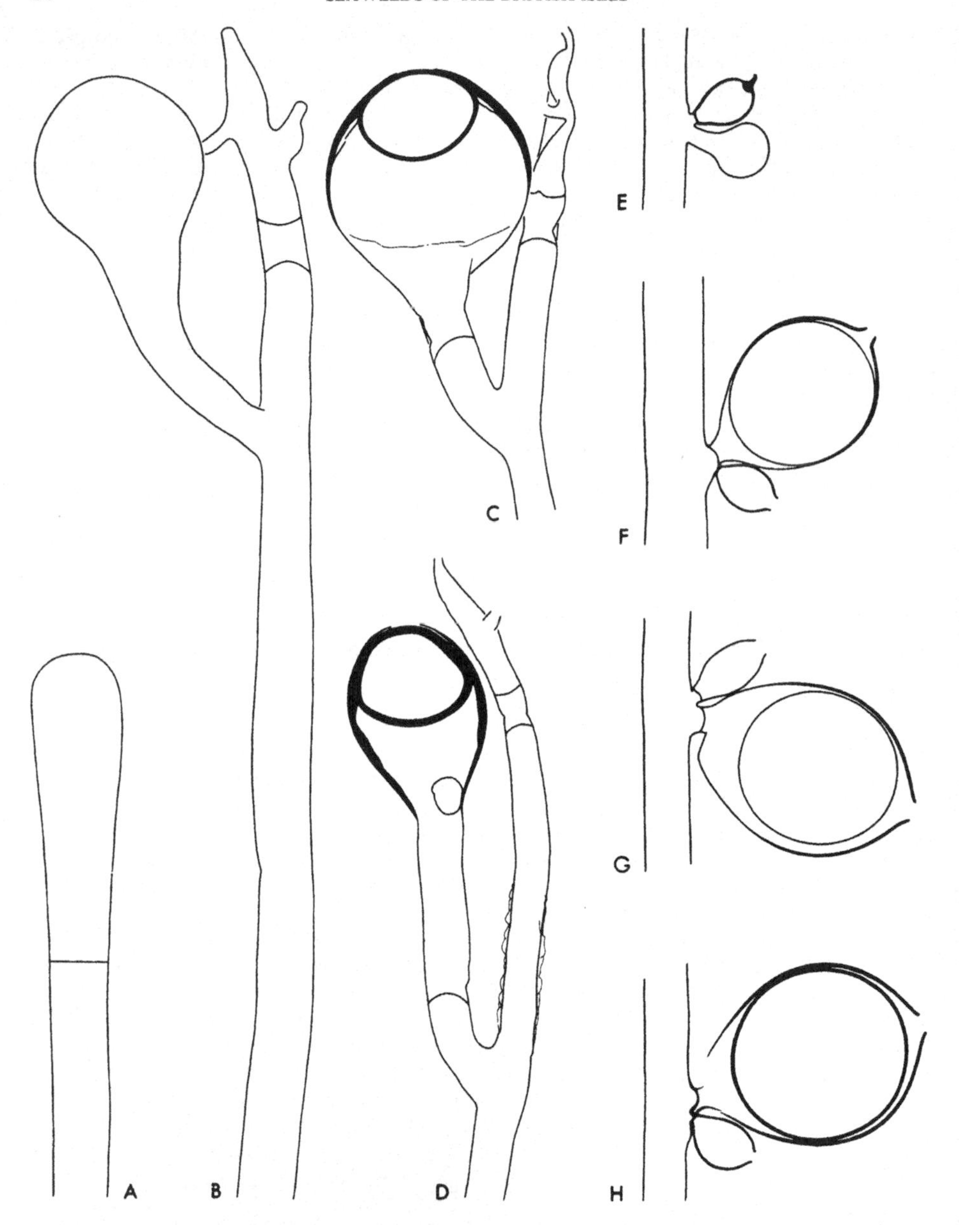

Fig. 6. A–D. *Vaucheria piloboloides*; A. Aplanosporangium; after Ernst (1904), non-British material; B–D. Sexual organs, in D also remains of what has probably been the wall of an aplanosporangium; E–H. *Vaucheria sescuplicaria*. All × 100.

This species is the only truly marine *Vaucheria* of the British flora. Capable of growing between 15 and 55‰ salinity but only growing well between 25 and 45‰ and fruiting between 25 and 35‰. Forming extensive growths on sandy bottoms in sheltered bays, mainly in the sublittoral, but occasionally also a little above low water mark.

Only known from Cornwall, South Devon, Dorset and Clare.

South-west Norway to the Mediterranean, Bermuda, Pakistan and India.

Two closely allied species have been described from outside Europe, *V. bermudensis* Taylor & Bernatowicz (1952) and *V. caloundrensis* Cribb (1960). The former has been reported from the aquarium at Roscoff, Brittany. The variation of *V. piloboloides* should be studied more closely in comparison with these two allies.

The species appears to occur at most times of the year in one place or another. Fruiting material has been found in June and September. Aplanospores are formed before the sexual organs on material brought into crude culture in the laboratory; there are no certain records of aplanospores in nature, but the wrinkled wall seen in Fig. 6D just above the point of branching is probably an old sporangium wall, and was drawn from material collected in June.

Vaucheria sescuplicaria Christensen (1952), p. 182.

Holotype: C. Denmark (Skarum, Mors).

Vaucheria dichotoma f. *arternensis* Rieth (1953), p. 337.

Plants most usually occurring as masses of relatively long, little branched, erect filaments in water, but also sometimes forming carpets of short, densely packed, upright filaments on soil. Filaments 69 ± 16 μm wide. Chloroplasts without pyrenoids, rather variable in shape, generally elongate, some of them being elliptic, some with bluntly pointed ends.

Antheridium lateral on the vegetative filament, shortly spindle-shaped with a very narrow base, the maximum diameter usually a little above the middle, at maturity opening by the dissolving of an apical papilla, which leaves a slightly raised rim around the pore. Oogonium arising from the vegetative filament right at the base of the antheridium, similar to it in shape but much larger and proportionally broader, for some time displaying a dark green cap sharply delimited from the pale lower part, opening like the antheridium by the dissolving of an apical papilla. Oospore approximately spherical, length more rarely distinctly greater than breadth, length 216 ± 18 μm, breadth 202 ± 17 μm; wall almost colourless, usually 2–3 μm thick; contents for some time appearing pale golden-yellow owing to the presence of scattered, green chloroplasts and brown granules between the lipid droplets, eventually greyish with peripheral aggregations of brown granules.

Asexual spores unknown.

Grows well between 5 and 35‰ salinity and may show slight growth right from fresh water to 60‰ salinity in laboratory experiments while fruiting between 5 and 50‰. In nature most commonly found at relatively low salinities, often in small bodies of water behind embankments or in the uppermost parts of salt marshes, but also above water under various phanerogams, in such places often associated with other species, in particular *V. erythrospora*, *V. compacta* and *V. intermedia*.

Found all along the south-east coast of England, more scattered over the rest of the British Isles.

Iceland to Tunisia.

Found throughout the year. Fruiting in summer and autumn.

Rieth (1954) gives a detailed and abundantly illustrated description of the developmental stages in this species, including observations on a number of abnormalities.

Vaucheria subsimplex Crouan frat. (1867), p. 133.

Provisional lectotype: original description, in the absence of material (see Christensen, 1973). France (Baie de Saint-Marc, Finistère).

Vaucheria sphaerospora Nordstedt (1878), p. 177.

Plants usually occurring as densely packed, erect filaments on soil, not infrequently almost covered by silt with only the tips exposed, but also often forming velvety carpets or cushions or irregularly tufted masses. Filaments 38 ± 4 μm wide. Chloroplasts with pyrenoids, usually tapering to a long point at one or both ends.

Antheridium terminal on a vegetative filament, separated from it by an empty space, more or less curved, generally bending about 90°, over its greater part as wide as the vegetative filament, towards the apex tapering into a pointed tip, where an exit pore is eventually formed, usually with a small number of lateral projections terminated by similar pores, very often two opposite to one another perpendicular to the plane of the curvature. Oogonium developing from the part of the filament immediately beneath the antheridium, starting as a saccate widening adjacent to the empty space, after which a double wall is formed at some distance below this widening, separating the oogonium from the vegetative thallus. Contents of the oogonium dividing into a spherical egg situated in the bulge and a smaller portion of protoplasm located in the cylindrical part of the oogonium, the smaller portion soon dying, sometimes after forming a thin inner wall across the oogonium. Opening by the dissolving of part of the wall on the side away from the tube without previous formation of a papilla. Oospore approximately spherical, 129 ± 17 μm long in the direction of the bulge, 123 ± 16 μm across; wall about 4 μm thick, almost colourless; contents pale golden-yellow.

Aplanospores reported from the Netherlands (Simons, 1975a).

Capable of growing between 5 and 35‰ salinity. Fruiting between 10 and 30‰. Found in salt marshes, often on almost bare ground below the level of the closed phanerogam cover, otherwise most usually under *Halimione*, very often mixed with *V. velutina*, also quite often with *V. compacta* and *V. intermedia*.

Generally distributed throughout the British Isles.

Greenland to north-west France and Labrador.

Found throughout the year, fruiting from March to November.

Vaucheria synandra Woronin (1869), column 137.

Holotype: PC. France (Nice).

Plants usually forming a relatively thin, loose to dense covering on soil, but occasionally also rising into soft cushions. Filaments 51 ± 7 μm wide. Chloroplasts without pyrenoids, usually elongate with bluntly pointed ends, sometimes almost circular in outline.

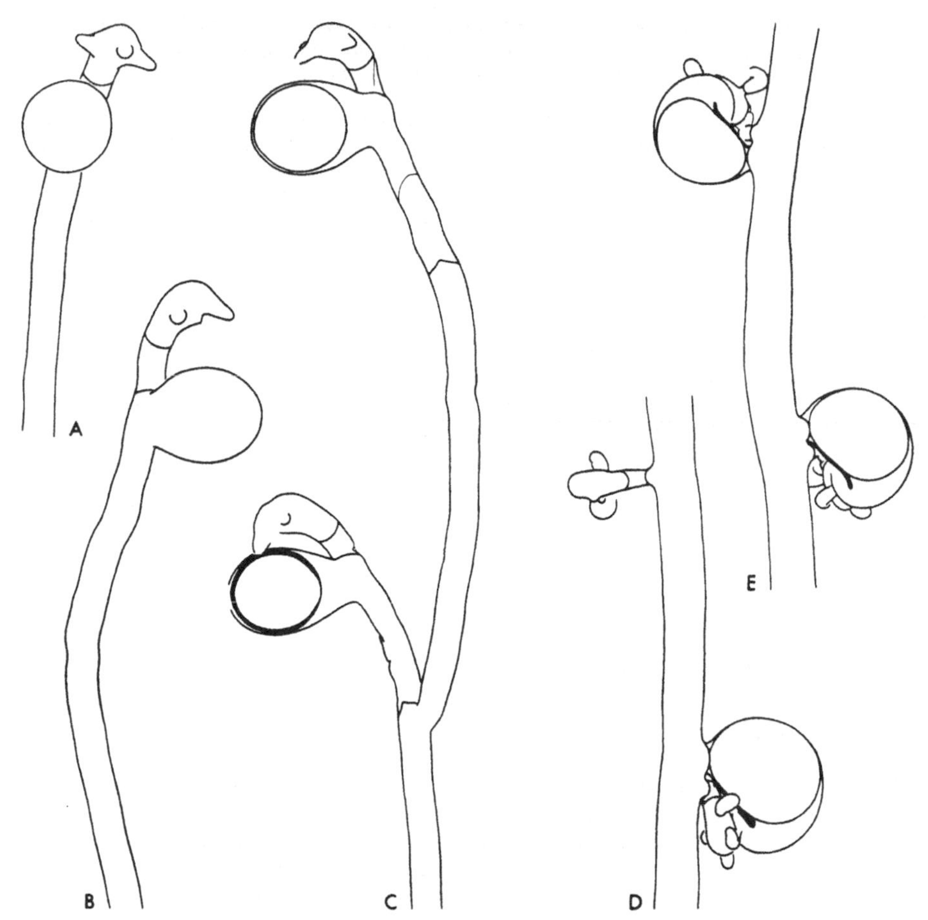

Fig. 7 A–C. *Vaucheria subsimplex*; D–E. *Vaucheria synandra*. All × 100.

Antheridia borne on a special, shortly cylindrical or clavate structure, the so-called androphore, which is a little narrower than the vegetative filament, growing out from it at a right angle and becoming separated from it by an empty space. Most usually 3 or 4 antheridia are developed on one androphore, projecting in all directions, each separated from it by a double wall, shortly tubular and pronouncedly curved, opening at the tip. Oogonium sessile on the main filament, growing out from it right at the base of a previously formed androphore, curved towards the androphore and thereby pushing it aside, opening at the tip of a hooked apical beak. Oospore leaving a considerable unfilled space at the beak, otherwise filling the oogonium, taking its shape and adhering to its wall except for a frequent slight rounding off at the base, 139 ± 14 µm long, 115 ± 11 µm across in lateral view; wall at maturity usually 4–8 µm thick with a thin, colourless outer

layer and on the inner side a much heavier golden-brown layer, which is often finely punctate; contents with one or more dark brown spots, otherwise almost colourless.
Asexual spores probably not formed (cf. Christensen, 1986a).

Capable of growing between 2·5 and 50‰ salinity and of fruiting from about the same lower limit up to 35‰. In nature mainly found at low salinities within reach of only the highest tides, under *Phragmites*, smaller grasses, riverside plants, or trees, but also not uncommonly in salt-marsh grassland, often growing alone, otherwise with other species that can tolerate low salinities well, mainly *V. compacta, V. coronata, V. medusa, V. sescuplicaria* and the halophilous freshwater species *V. canalicularis*.
Scattered finds from all over the British Isles.
Iceland to Algeria, Louisiana.

British finds from April to November, fruiting material only from June and November.
The British material of this species has somewhat narrower filaments and somewhat larger oospores than indicated in the original description. Liebetanz (1926, p. 113) has proposed a separate variety, var. *halophila,* for such plants. To consider the justification of such separation one should study the entire variation of the species.

Vaucheria velutina C. Agardh (1824), p. 312.

Holotype: LD. Sweden (Landskrona).

Vaucheria thuretii Woronin (1869), column 157.

Plants usually associated with other *Vaucheria* species, occurring as densely matted, upright filaments, almost covered by silt with only the tips visible, or forming velvet-like carpets, or cushions, or irregularly tufted. Filaments 55 ± 7 µm wide. Chloroplasts relatively broad with a pyrenoid at one edge, quite often lying with the pyrenoid towards the vacuole and a sharp edge towards the wall.
Antheridia sessile on the main filaments or borne on extremely short pedicels, generally 2–4 or more together, all of the same age, sack-shaped, most frequently more or less bent but also quite often upright, opening by the dissolving of an apical papilla. Oogonia sessile or shortly pedicellate like the antheridia, usually one with each group of antheridia, developing on the side furthest from the tip of the filament, pear-shaped with a broad connection to the filament (unlike in *V. sescuplicaria*), generally bent towards the antheridia but also sometimes erect, opening in the same way as the antheridia. Oospore approximately spherical, 166 ± 17 µm long, 161 ± 14 µm wide; wall about 3 µm thick, smooth, usually almost colourless; contents pale golden-yellow owing to the presence of a few green chloroplasts and some golden brown granules between the lipid droplets.
Aplanospores reported from the Netherlands (Nienhuis & Simons, 1971) and, with caution, from Massachusetts (Farlow, 1881).

Growing quite well in fresh water in the laboratory and a little even at a salinity as high as 60‰, but best between 5 and 10‰. Fruiting between 2·5 and 30‰. In nature this species has not been found in fresh water. It sometimes occurs at rather low salinities under *Phragmites, Scirpus maritimus*, etc. but is more frequently found right by the sea, on the sides of creeks, on mud flats and under *Halimione, Spartina, Salicornia* and *Puccinellia*.

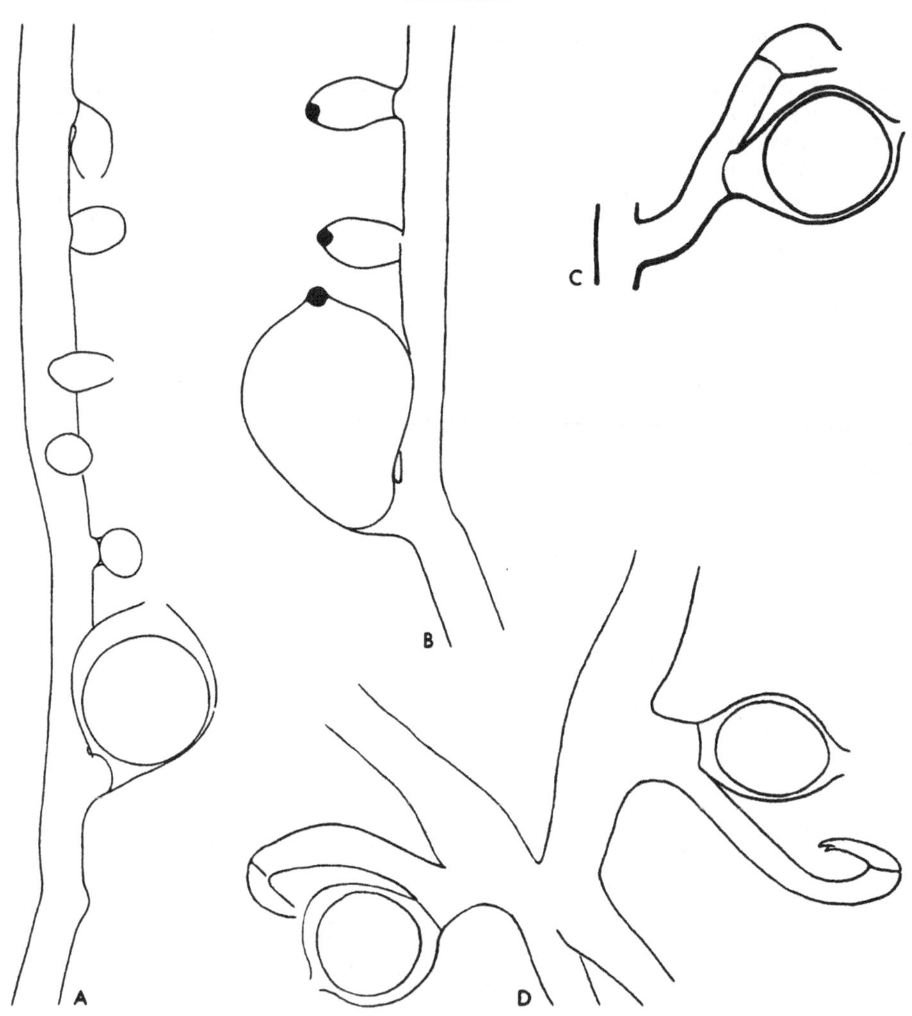

Fig. 8 A–B. *Vaucheria velutina*; C. *Vaucheria vipera*; after Yamagishi (1964), non-British material; D. Probably *Vaucheria vipera*; after Carter (1933). All × 100.

Nearly always mixed with other species, mainly *V. subsimplex, V. intermedia* and *V. compacta*. A single sample was collected in the sublittoral, and included no other species.

One of the commoner species, although collections are rather scattered in some areas.

Iceland to Algeria, Atlantic and Pacific coasts of North America, Pakistan and India, South Australia.

Found throughout the year, fruiting material from May to October.

Older reports must be regarded with reservation as samples of *V. sescuplicaria* have often been referred to *V. velutina*. Material with antheridia and oogonia on separate filaments and with most of the oogonia upright may perhaps be referred to var. *separata* Christensen (1986, p. 22) (syn. *V. submarina* sensu de Wildeman (1897), vix. *V. dichotoma* var. *submarina* Lyngbye = *V. submarina* (Lyngbye) Berkeley). A plant collected on several occasions from Weymouth by Berkeley and Holmes was assumed by these authors to be the same as that described by Lyngbye. As far as can be seen from the herbarium material most of its sexual organs are abortive, while the rest are like those of var. *separata*. A recent collection from the same area seems to vacillate between this variety and var. *velutina* (cf. Christensen, 1986).

Vaucheria vipera Blum (1960), p. 298.

Isotype: NY. USA (Barnstable, Mass.), cf. Christensen, 1986a.

?*Vaucheria woroniniana* sensu Carter (1933), p. 151 , non *Vaucheria woroniniana* Heering (1907), p. 165.

Filaments (22–) 50 (–100) µm wide.

Antheridium approximately tubular, sometimes slightly widened in the middle part, tapering towards the apex, more or less curved, rather variable in length, terminal on a main filament or a fruiting branch, separated from the vegetative tube by a double wall, opening at the apex. Oogonium sessile in a lateral position on the tube that carries the antheridium at its end, more or less pear-shaped with a slightly raised apex, where a pore is formed. Oospore approximately spherical, (97–) 125 (–190) µm in diameter, brownish.

Asexual spores unknown.

Found on salt marshes, often with *V. velutina*.

There are no authenticated finds of this species from Britain, but in the paper by Carter (*loc. cit.*) there is an illustration of material from Essex referred, with doubt, to *V. woroniniana*, and this illustration resembles the later described *V. vipera* more than anything else. Carter's drawing is reproduced here as Fig. 8D, and an illustration of material of *V. vipera* from Japan as Fig. 8C.

Netherlands, Massachusetts, Connecticut, Japan, Hong Kong.

Fruiting in August and September in the Netherlands.

REFERENCES FOR TRIBOPHYCEAE

ÅBERG. H. 1978. Light and branch formation in the alga *Vaucheria dichotoma* (Xanthophyceae). *Physiol. Plant.* **44:** 224–230.

ÅBERG, H. & FRIES, L. 1976. On cultivation of the alga *Vaucheria dichotoma* (Xanthophyceae) in axenic culture. *Phycologia* **15:** 133–141.

AGARDH, C. A. 1823. *Species Algarum* 1, pp. 399–531. Gryphiswaldiæ.

AGARDH, C. A. 1824. *Systema Algarum.* Lundæ.

BERKELEY, M. J. 1833. *Gleanings of British Algae.* London.

BLUM, J. L. 1960. A new *Vaucheria* from New England. *Trans. Am. micr. Soc.* **79:** 298–301.

BLUM, J. L. 1972. Vaucheriaceae, in *North American Flora,* series 2, part 8. New York.

BLUM, J. L. & CONOVER, J. T. 1953. New or noteworthy Vaucheriae from New England salt marshes. *Biol. Bull.* **105:** 395–401.

BLUM, J. L. & WILCE, R. T. 1958. Description, distribution and ecology of three species of *Vaucheria* previously unknown from North America. *Rhodora* **60:** 283–288.

BLUM, J. L. & WILCE, R. T. 1958a. The type of *Vaucheria compacta* var. *koksoakensis. Rhodora* **60:** 329.

BLUMENBACH, J. F. (1781). Über eine ungemein einfache Fortpflanzungsart. *Gött. Mag. Wiss. Litt.* **2:** 80–89.

BOHLIN, K. 1901. *Utkast till de gröna algernas och arkegoniaternas fylogeni.* Upsala.

CANDOLLE, A. P. de (le C. Decandolle) 1801. Extrait d'un rapport sur les Conferves fait à la Société philomathique. *Bull. Soc. philomath.* **3:** 17–21.

CARTER, N. 1933. A comparative study of the alga flora of two salt marshes. Part II. *J. Ecol.* **21:** 128–208.

CHAPMAN, V. J. 1937. A revision of the marine algae of Norfolk. *J. Linn. Soc., Bot.* **51:** 205–263.

CHRISTENSEN, T. 1952. Studies on the genus *Vaucheria* I. A list of finds from Denmark and England with notes on some submarine species. *Bot. Tidsskr.* **49:** 171–188.

CHRISTENSEN, T. 1956. Studies on the genus *Vaucheria* III. Remarks on some species from brackish water. *Bot. Not.* **109:** 275–280.

CHRISTENSEN, T. 1957. Three species of *Vaucheria* new to Britain. *Br. phycol. Bull.* **1**(5): 43.

CHRISTENSEN, T. 1968. *Vaucheria* types in the Dillenian herbaria. *Br. phycol. Bull.* **3:** 463–469.

CHRISTENSEN, T. 1973. Some early *Vaucheria* descriptions. *Bot. Not.* **126:** 513–518.

CHRISTENSEN, T. 1980. *Algae, a taxonomic survey.* Fasc. 1. Odense.

CHRISTENSEN, T. 1985. *Microspora ficulinae,* a green alga living in marine sponges. *Br. phycol. J.* **20:** 5–7.

CHRISTENSEN, T. 1986. On the identity of *Vaucheria submarina* auct. (Tribophyceae). *Br. phycol. J.* **21:** 19–23.

CHRISTENSEN, T. 1986a. Typification of the British salt- and brackish-water species of *Vaucheria* (Tribophyceae). *Br. phycol. J.* **21:** 275–280.

CHRISTENSEN, T. 1987. Salinity preference of twenty species of *Vaucheria* (Tribophyceae). In preparation.

COLLINS, F. S. 1900. Notes on Algae – II. *Rhodora* **2:** 11–14.

CORRENS, C. 1893. Ueber eine neue braune Süsswasseralge, *Naegeliella flagellifera* nov. gen. et sp. *Ber. dt. bot. Ges.* **10:** 629–636.

COTTON, A. D. 1912. Marine algae, *in* Praeger, R. L., A biological survey of Clare Island in the county of Mayo, Ireland and of the adjoining district. *Proc. R. Ir. Acad.* **31,** sect. 1(15): 1–178.

CRIBB, A. B. 1960. Records of marine algae from south-eastern Queensland V. *Pap. Dep. Bot. Univ. Qd* **4:** 1–31.

CROUAN, P. L. & H. M. 1867. *Florule du Finistère.* Paris & Brest.

CULLINANE, J. P. 1974. Identification of the marine species of the genus *Vaucheria* in Ireland. *Proc. R. Ir. Acad.* **74,** sect. B: 403–410.

DANGEARD, P. A. 1925. La structure des Vauchéries dans ses rapports avec la terminologie nouvelle des éléments cellulaires. *Cellule* **35:** 237–250.

DANGEARD, P. J. L. (P. Dangeard) 1939. Le genre *Vaucheria*, spécialement dans la région du sud-ouest de la France. *Botaniste* **29:** 183–265.

DE WILDEMAN, É. 1897. Observations sur les algues rapportées par M. J. Massart d'un voyage aux Indes Néerlandaises. *Annls. Jard. bot. Buitenz.* suppl. **1:** 32–106.

DILLENIUS, J. J. 1742 ('1741'). *Historia Muscorum*. Oxonii.

DOWIDAR, A. R. & ROBSON, T. O. 1972. Studies on the biology and control of *Vaucheria dichotoma* found in freshwaters in Britain. *Weed Res.* **12:** 221–228.

DUMORTIER, B.-C. 1822. *Commentationes botanicæ. Observations botaniques*. Tournay.

ERNST, A. 1904. Siphoneen-Studien. III. Zur Morphologie und Physiologie der Fortpflanzungszellen der Gattung *Vaucheria* DC. 1. Sporangien- und Aplanosporangienbildung bei *Vaucheria piloboloides* Thur. *Beih. Bot. Centralbl.* **16:** 367–382.

FARLOW, W. G. 1881. The marine algae of New England, pp. 1–210 in *United States Commission of Fish and Fisheries. Report of the commissioner for 1879. Appendix A. Natural History. The miscellaneous documents of the senate of the United States for the second session of the forty-sixth congress, 1879–80*. [Also as reprint: *The marine algae of New England and adjacent coast*].

FELDMANN, J. 1941. Une nouvelle Xanthophycée marine: *Tribonema marinum* nov. sp. *Bull. Soc. Hist. nat. Afr. N.* **32:** 56–61.

FRITSCH, F. E. 1935. *The structure and reproduction of the Algae*. 1. Cambridge.

GAILLON, B. 1833. *Aperçu d'histoire naturelle et observations sur les limites qui séparent le règne végétal du règne animal*. Boulogne-sur-Mer. [Four different printings from the same year, cf. *Croman* in *Taxon* **8:** 60–61, 1959].

GÖTZ, H. 1897. Zur Systematik der Gattung *Vaucheria* DC. speciell der Arten der Umgebung Basels. *Flora* **83:** 88–134. [Also as dissertation with separate pagination. München].

HABEEB, H. 1965. *Debsalga gigasporangia* new genus and species of Vaucheriaceae. *Leaflets Acad. Biol.* **38:** 1–2.

HANSEN, G. 1972. *Oogenese hos Vaucheria sescuplicaria Chr.* Unpublished thesis, University of Copenhagen.

HEERING, W. 1907. Die Süßwasseralgen Schleswig-Holsteins. 2. Teil: Chlorophyceae (Allgemeines.– Siphonales). *Jb. hamb. wiss. Anst.* **24,** Beih. **3:** 103–235.

HEERING, W. 1921. Chlorophyceae IV. Siphonocladiales, Siphonales, *in* Pascher, A., *Die Süsswasser-Flora Deutschlands, Österreichs und der Schweiz* **7.** Jena.

HEIDINGER, W. 1908. Die Entwicklung der Sexualorgane bei *Vaucheria*. *Ber. dt. bot. Ges.* **26:** 313–363.

HIBBERD, D. J. 1981. Notes on the taxonomy and nomenclature of the algal classes Eustigmatophyceae and Tribophyceae (synonym Xanthophyceae). *J. Linn. Soc., Bot.* **82:** 93–119.

HOLMES, E. M. 1886. British marine algæ. *Scott. Nat.* **1886:** 258–264.

KNUTZEN, J. 1973. Marine species of *Vaucheria* (Xanthophyceae) in South Norway. *Norw. J. Bot.* **20:** 163–181.

KÜTZING, F. T. 1843. *Phycologia generalis oder Anatomie, Physiologie und Systemkunde der Tange*. Leipzig.

KÜTZING, F. T. 1856. *Tabulae Phycologicae*. **6.** Nordhausen.

LAMARCK, J.-B. P. A. de Monet, chevalier de & CANDOLLE, A. P. de (de Lamarck & Decandolle) 1805. *Flore française*, ed. 3, vol. 2. Paris.

LE JOLIS, A. 1863. *Liste des algues marines de Cherbourg*. Paris & Cherbourg.

LIEBETANZ, B. 1926. Hydrobiologische Studien an Kujawischen Brackwässern. *Bull. int. Acad. pol. Sci. Lett.*, Classe Sci. math. nat., Série B. Sci. nat., I. Bot. **1925:** 1–116.

LINNAEUS, C. 1753. *Species Plantarum*, 2. Holmiæ.

LUTHER, A. 1899. Ueber *Chlorosaccus* eine neue Gattung der Süsswasseralgen, nebst einigen Bemerkungen zur Systematik verwandter Algen. *Bih. K. svenska VetenskAkad. Handl.* **24,** Afd. 3, No. 13: 1–22.

LUTHER, H. 1953. *Vaucheria Schleicheri* de Wild. neu für Nordeuropa. *Memo. Soc. Fauna Flora fenn.* **28:** 32–40.

LYNGBYE, H. C. 1819. *Tentamen hydrophytologiae Danicae*. Hafniae.

MARTIUS, C. F. P. 1817. *Flora cryptogamica Erlangensis*. Norimbergae.

MOESTRUP, Ø. 1970. On the fine structure of the spermatozoids of *Vaucheria sescuplicaria* and on the later stages in spermatogenesis. *J. mar. biol. Ass. U.K.* **50:** 513–523.

MÜLLER, O. F. 1779. Von unsichtbaren Wassermosen. *Beschäft. berl. Ges. naturf. Freunde* **4:** 42–54.

MÜLLER, O. F. 1788. De Confervis palustribus oculo nudo invisibilibus. *Nova Acta Acad. Sc. imp. Petrop.* **3:** 89–98.

NIENHUIS, P. H. & SIMONS, J. 1971. *Vaucheria* species and some other algae on a Dutch salt marsh, with ecological notes on their periodicity. *Acta Bot. Neerl.* **20:** 107–118.

NORDSTEDT, C. F. O. 1878. Algologiska småsaker. 1. *Bot. Not.* **1878:** 176–180.

NORDSTEDT, C. F. O. 1879. Algologiska småsaker. 2: *Vaucheria*-studier 1879. *Bot. Not.* **1879:** 177–190.

NORDSTEDT, C. F. O. 1886. Some remarks on British submarine *Vaucheriae*. *Scott. Nat.* **1886:** 382–384.

OLTMANNS, F. 1895. Ueber die Entwickelung der Sexualorgane bei *Vaucheria*. *Flora* **80:** 388–420.

OTT, D. W. & BROWN, R. M. 1972. Light and electron microscopical observations on mitosis in *Vaucheria litorea* Hofman ex C. Agardh. *Br. phycol. J.* **7:** 361–374.

OTT, D. W. & BROWN, R. M. 1978. Developmental cytology of the genus *Vaucheria* IV. Spermatogenesis. *Br. phycol. J.* **13:** 69–85.

OTT, D. W. & HOMMERSAND, M. H. 1974. *Vaucheriae* of North Carolina. I. Marine and brackish water species. *J. Phycol.* **10:** 373–385.

PASCHER, A. 1912. Zur Gliederung der Heterokonten. *Hedwigia* **53:** 6–22.

POLDERMAN, P. J. G. 1974. Some notes on three algae of salt-marshes on the east coast of England. *Br. phycol. J.* **9:** 251–253.

POLDERMAN, P. J. G. 1978. Algae of saltmarshes on the south and southwest coasts of England. *Br. phycol. J.* **13:** 235–240.

POLDERMAN, P. J. G. 1979. The saltmarsh algal communities in the Wadden area, with reference to their distribution and ecology in N.W. Europe. I. The distribution and ecology of the algal communities. *J. Biogeogr.* **6:** 225–266.

POLDERMAN, P. J. G. 1980. The saltmarsh algal communities in the Wadden area, with reference to their distribution and ecology in N.W. Europe. II. The zonation of algal communities in the Wadden area. *J. Biogeogr.* **7:** 85–95.

POLDERMAN, P. J. G. 1980a. The saltmarsh algal communities in the Wadden area, with reference to their distribution and ecology in N.W. Europe. III. The classificatory and semantic problems of saltmarsh algal communities. *J. Biogeogr.* **7:** 115–126.

POLDERMAN, P. J. G. & POLDERMAN-HALL, R. A. 1980. Algal communities in Scottish saltmarshes. *Br. phycol. J.* **15:** 59–71.

PRINGSHEIM, N. 1855. Über die Befruchtung der Algen. *Ber. Verh. K. preuss. Akad. Wiss. Berl.* **1855:** 133–165.

RIETH, A. 1953. Eine neue *Vaucheria* der Sektion *Woroninia* aus dem Arterner Salzgebeit. *Arch. Protistenk.* **98:** 327–341.

RIETH, A. 1954. Beobachtungen zur Entwicklungsgeschichte einer *Vaucheria* der Sektion *Woroninia*. *Flora* **142:** 156–182.

RIETH, A. 1956. Zur Kenntnis halophiler Vaucherien. *Flora* **143:** 127–160.

RIETH, A. 1956a. Beitrag zur Kenntnis der Phycomyceten III. *Kulturpflanze* **4:** 181–186.

RIETH, A. 1978. Beiträge zur Kenntnis der Vaucheriaceae XXI. Monözie und Diözie im Formenkreis von *Vaucheria dichotoma* (L.) AGARDH und die Art *Vaucheria starmachii* Kadłubowska. *Arch. Protistenk.* **120:** 409–419.

RIETH, A. 1980. Xanthophyceae, 2. Teil, *in* Pascher, A., *Süsswasserflora von Mitteleuropa* **4.** Stuttgart & New York.

ROSENVINGE, L. K. 1879. *Vaucheria sphærospora* Nordst. v. *dioica* n. var. *Bot. Not.* **1879:** 190.

ROTH, A. W. (A. G. Roth). 1797. *Catalecta Botanica* 1. Lipsiae.

SILVA, P. C. 1980. *Names of classes and families of living algae.* Regnum Vegetabile vol. **103.** Utrecht & The Hague.

SIMONS, J. 1974. *Vaucheria compacta*: a euryhaline estuarine algal species. *Acta bot. Neerl.* **23:** 613–626.

SIMONS, J. 1975. *Vaucheria* species from estuarine areas in the Netherlands. *Netherl. Sea Res.* **9:** 1–23.

SIMONS, J. 1975a. Periodicity and distribution of brackish *Vaucheria* species from non-tidal coastal areas in the S.W. Netherlands. *Acta Bot. Neerl.* **24:** 89–110.

SIMONS, J. & VROMAN, M. 1968. Some remarks on the genus *Vaucheria* in the Netherlands. *Acta Bot. Neerl.* **17:** 461–466.

SOLMS-LAUBACH, H. ZU 1867. Ueber *Vaucheria dichotoma* DC. *Bot. Zeitung* **25:** 361–366.

STARMACH, K. 1972. Chlorophyta III. *In:* Starmach, K. & Siemińska, J. *Flora słodkowodna Polski* **10.** Kraków.

TAYLOR, W. R. 1937. Notes on North Atlantic marine algae. I. *Pap. Mich. Acad. Sci.* **22:** 225–233.

TAYLOR, W. R. & BERNATOWICZ, A. J. 1952. Bermudian marine Vaucherias of the section Piloboloideae. *Pap. Mich. Acad. Sci.* **37:** 75–85.

THURET, G.-A. 1854. Sur quelques algues nouvelles. *Mém. Soc. Sci. nat. Cherbourg* **2:** 387–389.

TRENTEPOHL, J. F. 1807. Beobachtungen über die Fortpflanzung der Ectospermen des Herrn Vaucher insonderheit der *Conferva bullosa* Linn. nebst einigen Bemerkungen über die Oscillatorien, pp. 180–216 *in* Roth. A. W. (ed.) *Botanische Bemerkungen und Berichtigungen.* Leipzig.

VAUCHER, J.-P.-E. (J.-P. Vaucher) 1803. *Histoire des conferves d'eau douce.* Genève.

VENKATARAMAN, G. S. 1961. *Vaucheriaceae.* New Delhi.

WALZ, J. 1866. Beitrag zur Morphologie und Systematik der Gattung *Vaucheria* DC. *(Pringsh.) Jb. wiss. Bot.* **5:** 127–160.

WHITFORD, L. A. 1943. The fresh-water Algae of North Carolina. *J. Elisha Mitchell scient. Soc.* **59:** 131–170.

WORONIN, M. 1869. Beitrag zur Kenntnis der Vaucherien. *Bot. Zeitung* **27:** 137–144, 153–161.

YAMAGISHI, T. 1964. Observations on some siphonaceous algae collected from Okinawa. *J. Jap. Bot.* **39:** 82–90.

YAMADA, Y. 1934. The marine Chlorophyceae from Ryukyu, especially from the vicinity of Nawa. *J. Fac. Sci. Hokkaido Univ.*, Ser. 5 Bot. **3:** 33–88.

ORIGIN OF MATERIAL USED FOR ILLUSTRATIONS

FIG. 2A,B,C: Carew, Pembroke.
FIG. 2D,F,G: Kidwelly, Carmarthen.
FIG. 2E: Milton, Muir of Ord, Ross and Cromarthy.
FIG. 3A: Penclawdd, Glamorgan.
FIG. 3B,E,F: Sutton Bridge, Lincoln.
FIG. 3C,D: East Sleekburn, Blyth, Northumberland.
FIG. 3G: Sheeps Green, Cambridge.
FIG. 3H,I,J: Adlingfleet, York.
FIG. 4A,B,C,D: Carse of Ardersier, Inverness.
FIG. 4E: Four Mile Bridge, Anglesey.
FIG. 4F,G: Freiston Shore, Boston, Lincoln.
FIG. 4H: Penclawdd, Glamorgan.
FIG. 4I: Kidwelly, Carmarthen.
FIG. 5A,B,C: Coombe, Bude, Devon.
FIG. 5D,E,F: Dollymound, Dublin.
FIG. 5G: Langstone Bridge, Hayling Island, Hampshire.
FIG. 6B,C,D: Falmouth Harbour, Cornwall.
FIG. 6E: Tràigh Ghriais, Lewis, Outer Hebrides.
FIG. 6F,G: Freiston Shore, Boston, Lincoln.
FIG. 6H: Lion Wharf, Canewdon, Essex.
FIG. 7A,B,C: Landimore, Cheriton, Glamorgan.
FIG. 7D,E: Old Mill Creek, Dartmouth, Devon.
FIG. 8A,B: Lion Wharf, Canewdon, Essex.

INDEX TO GENERA AND SPECIES

Synonyms in *italics*

CPSIA information can be obtained at www.ICGtesting.com
Printed in the USA
BVOW07s0605030615

402948BV00008B/39/P